Integral Integrals

Nate Ramer

December 2024 (revised November 2025)

Published by Amazon KDP

© 2025 by Nathan Ramer

Cover design copyright ©2024 by Nathan Ramer

All rights reserved. No part of this book may be scanned, uploaded, reproduced, distributed, or transmitted in any form or by any means whatsoever without written permission from the author, except in the case of brief quotations embodied in critical articles and reviews. Thank you for supporting the author's rights. Published 2024. Second edition published 2025.

"Derivatives are a science, integration is an art."

Contents

	0.1	Introduction	1
1	**Basic Techniques**		**3**
	1.1	U-Substitutions	3
	1.2	Integration by Parts	9
	1.3	Trigonometric Substitutions	14
	1.4	Partial Fraction Decomposition	19
2	**Substitutions**		**35**
	2.1	Weierstrauss Substitution	36
	2.2	Euler Substitutions	40
	2.3	Hyperbolic Trig Substitution	44
	2.4	Ninja Substitution	47
	2.5	Inversion	50
	2.6	King's Rule	54
	2.7	Queen's Rule	58
3	**More Techniques**		**61**
	3.1	Feynman's Technique	62
	3.2	"Sum" Crazy Technique	72
	3.3	Ramanujan's Master Theorem	78
	3.4	Frullani's Integrals	83

	3.5 Lobachevsky's Rule	88
	3.6 Polar Coordinates	90
4	**Special Functions**	**93**
	4.1 Gamma Function	94
	4.2 Beta Function	97
	4.3 Polygamma Function	101
	4.4 Zeta Function	107
5	**Contour Integration**	**113**
	5.1 Basics of Complex Analysis	114
	5.2 Contour Integrals	120
6	**Section on Series**	**131**
	6.1 Recognizing Functions	131
	6.2 Derivatives and Integrals	132
	6.3 Fourier Series	134
	6.4 Complex Analysis	136
7	**Applications**	**139**
	7.1 Physics	139
	7.2 Differential Equations	142
	7.3 Statistics	143
8	**Additional Problems**	**145**

0.1 Introduction

If you are reading this introduction, then it must be at least 2025 or later. I decided, after writing *Differential Equations*, that I needed to go back and revamp *Integral Integrals* and *Some Sums*. Once I was able to format my books in an aesthetically pleasing way, I was unstoppable. As of writing this, November 13th, 2025, I have sold 505 copies of my books. I never thought I would be able to do this, only first publishing *Integral Integrals* on December 27th, 2024. I am beyond ecstatic and feel my readers deserve the updated formatting. So, for this book, I tried to preserve as much of the original writing as possible. The only additions or subtractions are to add further clarity, remove redundancy, and improve formatting. Additionally, I added new exercises in at the end that I feel better suit the content. So, let us get straight into the second edition of *Integral Integrals*.

Chapter 1

Basic Techniques

This chapter covers the introductory techniques. Many of these concepts you may learn in a standard calculus class. For most integrals, these four methods will be able to solve the majority of problems; however, there are exceptions.

1.1 U-Substitutions

First, we can look at u-substitutions. It is also known as a "u-sub", and the "reverse chain rule". It is beneficial to simplify an integral by making use of a substitution.

> **Theorem 1.1.1: U-Substitution**
> $$\int_a^b f(g(x))g'(x)\ dx = \int_{g(a)}^{g(b)} f(u)\ du$$

In order to prove this theorem, we let F be the antiderivative of $f(x)$. By the chain rule,

$$F(g(x))' = F'(g(x))g'(x) \tag{1.1}$$

Since $F'(x) = f(x)$

$$F(g(x))' = f(g(x))g'(x) \tag{1.2}$$

Now, integrate both sides of the equation.

$$\int_a^b F(g(x))' \, dx = \int_a^b f(g(x))g'(x) \, dx \tag{1.3}$$

Applying the Fundamental Theorem of Calculus,

$$= F(g(b)) - F(g(a)) \tag{1.4}$$

Applying it again, with $u = g(x)$

$$\int_a^b f(g(x))g'(x) \, dx = \int_{g(a)}^{g(b)} f(u) \, du \tag{1.5}$$

This is the u-substitution in a nutshell. Although it is one thing to see the proof, it is another to know how to do it, so let's look at a couple of examples.

CHAPTER 1. BASIC TECHNIQUES

Example 1.1.1.

$$\int_4^5 14(x-5)^6 \, dx \tag{1.6}$$

This can also be solved by expanding the polynomial and integrating term by term using the power rule; however, as the power increases, this becomes increasingly unreasonable. Therefore, we use the u-substitution. In practice, you choose a function of x to be "u", take the derivative of it, plug it back in, and integrate the new function.
Let

$$u = x - 5 \tag{1.7}$$

Take the derivative of both sides with respect to x,

$$du = dx \tag{1.8}$$

Plug these back into the integral, with the new bounds

$$\int_{-1}^0 14u^6 \, du \tag{1.9}$$

See how much simpler it made things! Now, we can use the power rule:

$$= 2u^7 \, \big|_{-1}^0 \tag{1.10}$$

Giving the final answer,

$$\int_4^5 14(x-5)^6 \, dx = 2 \tag{1.11}$$

Example 1.1.2.

$$\int \frac{7}{(x-10)^7}\, dx \qquad (1.12)$$

Let
$$u = x - 10 \qquad (1.13)$$
$$du = dx \qquad (1.14)$$

Since the integral is indefinite, the bounds don't have to be changed, but the original function will have to be plugged back in at the end of the problem.

$$= \int 7u^{-7}\, du \qquad (1.15)$$

$$= \frac{-7}{6} u^{-6} \qquad (1.16)$$

Plugging back in u gives the final answer,

$$\int \frac{7}{(x-10)^7}\, dx = \frac{-7}{6}(x-10)^{-6} + C \qquad (1.17)$$

CHAPTER 1. BASIC TECHNIQUES

Example 1.1.3.

$$\int \frac{x^2 - 3}{-x^3 + 9x + 1} \, dx \qquad (1.18)$$

This one may not seem clear at first as it requires some manipulation after finding a u.
Let

$$u = -x^3 + 9x + 1 \qquad (1.19)$$

$$du = -3x^2 + 9 \, dx \qquad (1.20)$$

You may be thinking, "But this doesn't match anything else in the integral?". Well, if you divide both sides of that equation by -3, you arrive at

$$\frac{-1}{3} du = x^2 - 3 \, dx \qquad (1.21)$$

which is exactly what is in the original integral. Plugging this in turns the problem into

$$\frac{-1}{3} \int \frac{1}{u} \, du \qquad (1.22)$$

This is just the natural logarithm of the absolute value of u. Substituting the original function for u gives the final answer,

$$\int \frac{x^2 - 3}{-x^3 + 9x + 1} \, dx = \frac{-1}{3} \ln|-x^3 + 9x + 1| + C \qquad (1.23)$$

Example 1.1.4.

$$\int x\cos(2x^2)\,dx \qquad (1.24)$$

Let

$$u = 2x^2 \qquad (1.25)$$

$$du = 4x\,dx \qquad (1.26)$$

$$\frac{1}{4}du = x\,dx \qquad (1.27)$$

Plugging this into the integral gives

$$\frac{1}{4}\int \cos(u)\,du \qquad (1.28)$$

Integrating and plugging u back in gives the final answer,

$$\int x\cos(2x^2)\,dx = \frac{1}{4}\sin(2x^2) + C \qquad (1.29)$$

1.2 Integration by Parts

Integration by Parts, or "reverse product rule" is the second major basic integration type. It results when you take the integral of two functions multiplied together, where one of those is not the derivative of the other. The formula is given by

> **Theorem 1.2.1: Integration by Parts**
>
> $$\int u(x) \, dv(x) = uv - \int v \, du \qquad (1.30)$$

We construct a proof by using the product rule for derivatives.

$$\int \frac{d}{dx}(uv) \, dx = \int u \frac{dv}{dx} \, dx + \int v \frac{du}{dx} \, dx \qquad (1.31)$$

Canceling out the dx terms,

$$uv = \int u \, dv + \int v \, du \qquad (1.32)$$

Solving for $\int u \, dv$ gives the formula,

$$\int u \, dv = uv - \int v \, du \qquad (1.33)$$

In case you are struggling to remember the order of the variables, just remember "ultraviolet voodoo".

In practice, you will pick one function to take the derivative of, u, and another function dv to integrate.

Example 1.2.1.

$$\int xe^x \, dx \qquad (1.34)$$

Let

$$u = x \qquad (1.35)$$
$$du = dx \qquad (1.36)$$
$$dv = e^x \qquad (1.37)$$
$$v = e^x \qquad (1.38)$$

Now, plug these into the formula,

$$\int xe^x \, dx = xe^x - \int e^x \, dx \qquad (1.39)$$

Therefore,

$$\int xe^x \, dx = xe^x - e^x + C \qquad (1.40)$$

CHAPTER 1. BASIC TECHNIQUES

Example

$$\int e^x \cos x \, dx \qquad (1.41)$$

Let
$$u = e^x \qquad (1.42)$$
$$du = e^x \, dx \qquad (1.43)$$
$$dv = \sin x \qquad (1.44)$$
$$v = -\cos x \qquad (1.45)$$

Plug this into the formula,

$$\int e^x \cos x \, dx = e^x \sin x + e^x \sin x - \int e^x \cos x \, dx \qquad (1.46)$$

Since the integral on the left and right sides are the same, we can add them to the left side, and divide by two to get the final answer,

$$\int e^x \cos x \, dx = \frac{1}{2}(e^x \sin x + e^x \cos x) + C \qquad (1.47)$$

Example 1.2.2.

$$\int x^2 \sin x \, dx \qquad (1.48)$$

Let

$$u = x^2 \qquad (1.49)$$
$$du = 2x \, dx \qquad (1.50)$$
$$dv = \sin x \qquad (1.51)$$
$$v = -\cos x \qquad (1.52)$$

Plugging it into the formula,

$$= -x^2 \cos x + 2 \int x \cos x \, dx \qquad (1.53)$$

Now, we have to use integration by parts again since the result of doing it the first time is still a product of two functions. So, let

$$u = x \qquad (1.54)$$
$$du = dx \qquad (1.55)$$
$$dv = \cos x \qquad (1.56)$$
$$v = \sin x \qquad (1.57)$$

Plugging this in and combining with the previous result gives,

$$= -x^2 \cos x + 2x \sin x - 2 \int \sin x \, dx \qquad (1.58)$$

Integrating the last term gives the final answer,

$$\int x^2 \sin x \, dx = -x^2 \cos x + 2x \sin x + 2 \cos x + C \qquad (1.59)$$

CHAPTER 1. BASIC TECHNIQUES

Example 1.2.3.

$$\int \arctan x \, dx \tag{1.60}$$

For this one, let

$$u = \arctan x \tag{1.61}$$

$$du = \frac{1}{1+x^2} \, dx \tag{1.62}$$

$$dv = 1 \tag{1.63}$$

$$v = x \, dx \tag{1.64}$$

Plugging this in gives,

$$x \arctan x - \int \frac{x}{1+x^2} \, dx \tag{1.65}$$

Now, for the second integral, we have to do a regular u-sub. Let

$$u = 1 + x^2 \tag{1.66}$$

$$\frac{1}{2} du = x \, dx \tag{1.67}$$

The second integral then becomes,

$$\frac{1}{2} \int \frac{1}{u} \, du \tag{1.68}$$

Integrating, plugging u back in, and plugging that integral into the original result gives the final answer to the integral.

$$\int \arctan x \, dx = x \arctan x - \frac{1}{2} \ln(1+x^2) + C \tag{1.69}$$

1.3 Trigonometric Substitutions

Trigonometric substitutions are primarily useful when dealing with quadratics beneath square roots. Then, you can utilize the Pythagorean trigonometric identities to simplify them. Technically, these are just a specific example of a u-substitution, but they are useful enough to be taught separately.

$$\sin^2 x + \cos^2 x = 1 \tag{1.70}$$

$$\tan^2 x + 1 = \sec^2 x \tag{1.71}$$

$$1 + \cot^2 x = \csc^2 x \tag{1.72}$$

These three equations can be solved to find any of the six trigonometric functions. These are often the results of trigonometric substitutions. It works in the same way that a regular u-substitution does, except that usually, the variable of choice is θ.

There is a general rule for when to substitute which trigonometric functions where.
If the integral contains,

$$\sqrt{a^2 - x^2} \tag{1.73}$$

Then, substitute

$$x = a \sin \theta \tag{1.74}$$

$$\Rightarrow \sqrt{a^2 - x^2} = a \cos \theta \tag{1.75}$$

If the integral contains,

$$\sqrt{a^2 + x^2} \tag{1.76}$$

Then, substitute
$$x = a\tan\theta \tag{1.77}$$
$$\Rightarrow \sqrt{a^2 + x^2} = a\sec\theta \tag{1.78}$$

And if the integral contains,
$$\sqrt{x^2 - a^2} \tag{1.79}$$
Then, substitute
$$x = a\sec\theta \tag{1.80}$$
$$\Rightarrow \sqrt{x^2 - a^2} = a\tan\theta \tag{1.81}$$

1.3. TRIGONOMETRIC SUBSTITUTIONS

Example 1.3.1.

$$\int \frac{\sqrt{9-x^2}}{x^2} \, dx \qquad (1.82)$$

Substitute

$$x = 3\sin\theta \qquad (1.83)$$
$$dx = 3\cos\theta \qquad (1.84)$$
$$= \int \frac{3\cos\theta}{9\sin^2\theta} 3\cos\theta \, d\theta \qquad (1.85)$$

Simplifying and rewriting using the definition of $\cot\theta$,

$$= \int \cot^\theta \, d\theta \qquad (1.86)$$

Now, rewrite using the Pythagorean identity from above,

$$= \int \csc^2\theta - 1 \, d\theta \qquad (1.87)$$

$$= -\cot\theta - \theta \qquad (1.88)$$

Solve the original substitution for theta and substitute this in,

$$\theta = \arcsin\frac{x}{3} \qquad (1.89)$$

$$\Rightarrow -\cot\theta - \theta = -\cot\arcsin\frac{x}{3} - \arcsin\frac{x}{3} \qquad (1.90)$$

This, though technically correct, can be further simplified by constructing a triangle in terms of theta from the substitution and finding $\cot\theta$. Doing this gives the final answer to the integral,

$$\int \frac{\sqrt{9-x^2}}{x^2} \, dx = \frac{-\sqrt{9-x^2}}{x} - \arcsin\frac{x}{3} + C \qquad (1.91)$$

CHAPTER 1. BASIC TECHNIQUES

Example 1.3.2.

$$\int \frac{1}{x^3\sqrt{x^2-9}}\,dx \qquad (1.92)$$

Let

$$x = 2\sec\theta \qquad (1.93)$$

$$dx = 2\sec\theta\tan\theta\,d\theta \qquad (1.94)$$

Plugging this in,

$$\int \frac{2\sec\theta\tan\theta}{16\sec^3\theta\tan\theta}\,d\theta \qquad (1.95)$$

Simplifying,

$$\frac{1}{8}\int \sec^{-2}\theta\,d\theta \qquad (1.96)$$

This is also equal to

$$\frac{1}{8}\int \cos^2\theta\,d\theta \qquad (1.97)$$

Now, using the trigonometric identity $\cos^2\theta = \frac{1}{2}(1+\cos 2\theta)$, the integral can be written as,

$$\frac{1}{16}\int 1 + \cos 2\theta\,d\theta \qquad (1.98)$$

The second term of this is another u-sub, however, I feel you are the point that I can omit from showing this step. Integrating these terms gives,

$$\frac{1}{16}\theta + \frac{1}{32}\sin 2\theta \qquad (1.99)$$

1.3. TRIGONOMETRIC SUBSTITUTIONS

Another useful trig identity that allows us to rewrite the second term is $\sin 2\theta = 2\sin\theta\cos\theta$. Then, since all of the trig functions are in terms of just θ, you can solve for it in the substitution, substitute back in, and make the triangle. Doing all of this allows us to arrive at the final answer,

$$\int \frac{1}{x^3\sqrt{x^2-9}}\,dx = \frac{1}{16}\left(\sec^{-1}\frac{x}{2} + \frac{2\sqrt{x^2-4}}{x^2}\right) + C \quad (1.100)$$

1.4 Partial Fraction Decomposition

For our last major technique, we can look at what happens if you have ratios of polynomial functions. For example,

$$\int \frac{x^2 + 2x + 3}{x^3 - 2x^2 + 6x - 3} \, dx \qquad (1.101)$$

It can also have a highest degree of 0 on top, or the top can be 1, such as,

$$\int \frac{1}{1 - x^2} \, dx \qquad (1.102)$$

These problems come in two primary cases, which are described below.

Case 1

In this case, we are assuming that the highest degree of the polynomial in the numerator is greater than or equal to that of the polynomial in the denominator. For this, the first step is to divide the top polynomial by the bottom one. You can use either polynomial long division or synthetic division. Then, refer to the steps of case 2.

Case 2

In this case, the degree of the bottom is greater than that of the top. In this case, you will follow a series of steps.

1. Do the top and bottom factors cancel out like terms?

2. Is there a simple u-substitution that avoids using partial fractions?

3. Does the bottom factor at all? Are they linear, quadratic, or higher-order? Are they repeated factors?

These are all important questions to ask yourself as you evaluate these integrals; some of them may not even require partial fractions, despite being a ratio of polynomials. This is a rarity, however. So, we need to look at how to decompose the integrals into workable fractions.

CHAPTER 1. BASIC TECHNIQUES

The first, simplest case occurs when the denominator can be factored into a product of linear factors.

> **Definition 1.4.1: Non-repeating Linear Factors**
>
> f a rational function can be written in the form
>
> $$\frac{p(x)}{(x-x_0)(x-x_1)...} \qquad (1.103)$$
>
> It is said to contain linear non-repeating factors.

In order to decompose this, we rewrite it in the form

$$\frac{p(x)}{(x-x_0)(x-x_1)...} = \frac{A}{x-x_0} + \frac{B}{x-x_1} + ... \qquad (1.104)$$

Then, you would multiply $(x-x_0)(x-x_1)...$ on both sides and solve for the constants $A, B, ...$.

1.4. PARTIAL FRACTION DECOMPOSITION

> **Definition 1.4.2: Repeated Linear Factors**
>
> A rational function is said to contain repeated linear factors if it can be expressed in the form
>
> $$\frac{p(x)}{(x-x_0)^a(x-x_1)^b...} \qquad (1.105)$$
>
> Where $p(x)$ is an arbitrary polynomial, and $a, b, ...$ are natural numbers.

For example,

$$\frac{1}{(2x-1)^2} \qquad (1.106)$$

Has repeated linear factors; however,

$$\frac{1}{(2x+1)(x-2)(4x+3)} \qquad (1.107)$$

Does not because each of the terms in the denominator is distinct from one another.

Example 1.4.1.

$$\int \frac{x^4 - 2x^2 + 4x + 1}{x^3 - x^2 - x + 1} \, dx \qquad (1.108)$$

This requires long division before it can be decomposed. Doing this gives,

$$\int x + 1 + \frac{4x}{x^3 - x^2 - x + 1} \, dx \qquad (1.109)$$

Factor the denominator of the second term,

$$\int \frac{4x}{(x+1)(x-1)(x-1)} \qquad (1.110)$$

CHAPTER 1. BASIC TECHNIQUES

The decomposition setup is as follows:

$$\frac{4x}{(x-1)^2(x+1)} = \frac{A}{(x-1)^2} + \frac{B}{x-1} + \frac{C}{x+1} \quad (1.111)$$

Multiply everything by the denominator of the original fraction,

$$4x = A(x+1) + B(x-1)(x+1) + C(x-1)^2 \quad (1.112)$$

For this one, I will demonstrate a second method for determining the constants. Be careful, though; this only works with linear factors. What you do is plug in values that will "cancel out" certain constants. For example, plugging in $x = -1$ into the equation gives C

$$-4 = 4C \quad (1.113)$$

$$\Rightarrow C = -1 \quad (1.114)$$

Doing the same thing with $x = 1$ gives A,

$$4 = 2A \quad (1.115)$$

$$\Rightarrow A = 2 \quad (1.116)$$

Lastly, since we have run out of x-values that will cancel out terms, we can plug in an easy-to-evaluate number such as 0 and substitute our values of A and C to solve for B.

$$0 = A - B + C \quad (1.117)$$

$$0 = 2 - B - 1 \quad (1.118)$$

$$1 - B = 0 \quad (1.119)$$

1.4. PARTIAL FRACTION DECOMPOSITION

$$\Rightarrow B = 1 \qquad (1.120)$$

Now, rewrite the integral using the decomposed fraction

$$= \frac{1}{2}x^2 + x + 2\int \frac{1}{(x-1)^2}\,dx + \int \frac{1}{x-1}\,dx - \int \frac{1}{x+1}\,dx \qquad (1.121)$$

These three integrals are all u-substitutions and yield the final answer.

$$\int \frac{x^4 - 2x^2 + 4x + 1}{x^3 - x^2 - x + 1}\,dx = \frac{1}{2}x^2 + x - \frac{2}{x-1} + \ln\left|\frac{x-1}{x+1}\right| + C \qquad (1.122)$$

For the natural logarithm term, I combined it into one logarithm by using the fact that $\ln a - \ln b = \ln \frac{a}{b}$.

CHAPTER 1. BASIC TECHNIQUES

The next most complex case arises when you have non-factorable quadratics in the denominator.

> **Definition 1.4.3: Irreducible Quadratic Factors**
>
> A rational function has irreducible quadratic factors if it can be written in the form
> $$\frac{p(x)}{(a_0x^2 + b_0x + c_0)(a_1x^2 + b_1x + c_1)\ldots} \qquad (1.123)$$

In this case, the numerator of the decomposed fractions consists of $Ax + B$ instead of just a constant, A.

We can also combine the repeated aspects of linear factors with the complexity of the quadratic factors.

> **Definition 1.4.4: Repeated Irreducible Quadratics**
>
> rational function contains repeated irreducible quadratics if it can be written in the form
> $$\frac{p(x)}{(a_0x^2 + b_0x + c_0)^d(a_1x^2 + b_1x + c_1)^e} \qquad (1.124)$$

Example 1.4.2.

$$\int \frac{5x^2 + x - 3}{x^2 - 1}\, dx \qquad (1.125)$$

1.4. PARTIAL FRACTION DECOMPOSITION

Since the degrees of the numerator and denominator are equal. This problem first requires long division. Doing this gives

$$\int 5 + \frac{x+2}{x^2-1} \, dx \tag{1.126}$$

Now, the second integral falls into the second major case. The bottom factors down into $(x-1)(x+1)$, which does not perfectly cancel out with the top. We perform a partial fraction decomposition to break the integral down.

$$\frac{x+2}{(x+1)(x-1)} = \frac{A}{x+1} + \frac{B}{x-1} \tag{1.127}$$

A and B are both placeholder values. We will then solve for them, which will drastically simplify the integral that we have to evaluate. To do this, first, multiply by the denominator of the original ratio to eliminate all fractions.

$$x + 2 = A(x-1) + B(x-2) \tag{1.128}$$

Now, distribute the terms on the right-hand side.

$$x + 2 = Ax - A + Bx - 2B \tag{1.129}$$

Then, the factor is based on the degree of the attached polynomial. Group the x terms together and the constant terms together.

$$x + 2 = (A+B)x + (-A - 2B) \tag{1.130}$$

Set the coefficient of the x term on the left to the coefficient of the x term on the right.

$$1 = A + B \tag{1.131}$$

CHAPTER 1. BASIC TECHNIQUES

$$2 = -A - 2B \tag{1.132}$$

Solving this system of equations gives the values for A and B.

$$A = -\frac{1}{2} \tag{1.133}$$

$$B = \frac{3}{2} \tag{1.134}$$

Now that it has been split up into its partial fractions, the second term of the integral can be directly integrated using u-subs and natural logarithms.

$$\int 5 + \frac{x+2}{x^2-1}\, dx = 5x - \frac{1}{2}\ln|x+1| + \frac{3}{2}\ln|x-1| + C \tag{1.135}$$

This can be expanded to have as many terms as you'd like. You would just continue in the alphabet after A and B

1.4. PARTIAL FRACTION DECOMPOSITION

Example 1.4.3.

$$\int \frac{x^4 - 2x^2 + 4x + 1}{x^3 - x^2 - x + 1} \, dx \qquad (1.136)$$

This requires long division before it can be decomposed. Doing this gives,

$$\int x + 1 + \frac{4x}{x^3 - x^2 - x + 1} \, dx \qquad (1.137)$$

Factor the denominator of the second term,

$$\int \frac{4x}{(x+1)(x-1)(x-1)} \qquad (1.138)$$

The decomposition setup is as follows:

$$\frac{4x}{(x-1)^2(x+1)} = \frac{A}{(x-1)^2} + \frac{B}{x-1} + \frac{C}{x+1} \qquad (1.139)$$

Multiply everything by the denominator of the original fraction,

$$4x = A(x+1) + B(x-1)(x+1) + C(x-1)^2 \qquad (1.140)$$

For this one, I will demonstrate a second method for determining the constants. Be careful, though; this only works with linear factors. What you do is plug in values that will "cancel out" certain constants. For example, plugging in $x = -1$ into the equation gives C

$$-4 = 4C \qquad (1.141)$$

$$\Rightarrow C = -1 \qquad (1.142)$$

Doing the same thing with $x = 1$ gives A,

$$4 = 2A \tag{1.143}$$

$$\Rightarrow A = 2 \tag{1.144}$$

Lastly, since we have run out of x-values that will cancel out terms, we can plug in an easy-to-evaluate number such as 0 and substitute our values of A and C to solve for B.

$$0 = A - B + C \tag{1.145}$$

$$0 = 2 - B - 1 \tag{1.146}$$

$$1 - B = 0 \tag{1.147}$$

$$\Rightarrow B = 1 \tag{1.148}$$

Now, rewrite the integral using the decomposed fraction

$$= \frac{1}{2}x^2 + x + 2 \int \frac{1}{(x-1)^2} \, dx + \int \frac{1}{x-1} \, dx - \int \frac{1}{x+1} \, dx \tag{1.149}$$

These three integrals are all u-substitutions and yield the final answer.

$$\int \frac{x^4 - 2x^2 + 4x + 1}{x^3 - x^2 - x + 1} \, dx = \frac{1}{2}x^2 + x - \frac{2}{x-1} + \ln\left|\frac{x-1}{x+1}\right| + C \tag{1.150}$$

For the natural logarithm term, I combined it into one logarithm by using the fact that $\ln a - \ln b = \ln \frac{a}{b}$.

Example 1.4.4.

$$\int \frac{2x^2 - x + 4}{x(x^2 + 4)} \, dx \qquad (1.151)$$

Since the degree of the denominator is greater, and it is already factored, we can proceed directly with the partial fraction decomposition. For irreducible quadratic factors, it is set up as follows,

$$\frac{2x^2 - x + 4}{x(x^2 + 4)} = \frac{A}{x} + \frac{Bx + C}{x^2 + 4} \qquad (1.152)$$

Now, multiply by the denominator of the fraction,

$$2x^2 - x + 4 = A(x^2 + 4) + (Bx + C)(x) \qquad (1.153)$$

Distribute,

$$2x^2 - x + 4 = Ax^2 + 4A + Bx^2 + Cx \qquad (1.154)$$

Factor the right-hand side based on the degree of x attached,

$$2x^2 - x + 4 = (A + B)x^2 + (C)x + 4A \qquad (1.155)$$

Set the coefficients on the left-hand side to those on the right-hand side.

$$4A = 4 \qquad (1.156)$$
$$\Rightarrow A = 1 \qquad (1.157)$$
$$1 + B = 2 \qquad (1.158)$$
$$\Rightarrow B = 1 \qquad (1.159)$$
$$C = -1 \qquad (1.160)$$

CHAPTER 1. BASIC TECHNIQUES

Now, the original integral can be written as,

$$\int \frac{1}{x} + \frac{x-1}{x^2+4} \, dx \qquad (1.161)$$

The first integral is simply $\ln x$. For the second integral, split it up by the numerator.

$$\int \frac{x-1}{x^2+4} \, dx = \int \frac{x}{x^2+4} \, dx - \int \frac{1}{x^2+4} \, dx \qquad (1.162)$$

The first integral is a u-sub, and the second integral is a trig-sub. Solving these two yields the final answer,

$$\int \frac{2x^2 - x + 4}{x(x^2+4)} \, dx = \ln x + \frac{1}{2} \ln x^2 + 4 \qquad (1.163)$$

$$+ \frac{1}{2} \arctan \frac{x}{2} + C \qquad (1.164)$$

1.4. PARTIAL FRACTION DECOMPOSITION

Example 1.4.5.

$$\int \frac{1 - x + 2x^2 - x^3}{x(x^2+1)^2} \, dx \qquad (1.165)$$

Wow! What a hot mess! Let's set up the decomposition. Considering the last 3 types, this one is pretty intuitive.

$$\frac{1 - x + 2x^2 - x^3}{x(x^2+1)^2} = \frac{A}{x} + \frac{Bx+C}{(x^2+1)^2} + \frac{Dx+E}{x^2+1} \qquad (1.166)$$

Multiply everything out; you most likely know the drill by now.

$$1 - x + 2x^2 - x^3 = A(x^2+1)^2 + \qquad (1.167)$$
$$(Bx+C)(x) + (Dx+E)(x^2+1)(x) \qquad (1.168)$$
$$1 - x + 2x^2 - x^3 = Ax^4 + 2Ax^2 + A + Bx^2 + Cx \qquad (1.169)$$
$$+ Dx^4 + Dx^3 + Ex^3 + Ex \qquad (1.170)$$
$$1 - x + 2x^2 - x^3 = (A+D)x^4 + \qquad (1.171)$$
$$E(x^3) + (A+B+D)x^2 + (C+E)(x) + (A) \qquad (1.172)$$

$$A + D = 0 \qquad (1.173)$$
$$1 + D = 0 \qquad (1.174)$$
$$\Rightarrow A = 1 \qquad (1.175)$$
$$\Rightarrow D = -1 \qquad (1.176)$$
$$2A + B + D = 2 \qquad (1.177)$$
$$2 + B - 1 = 2 \qquad (1.178)$$
$$\Rightarrow B = 1 \qquad (1.179)$$
$$E = -1 \qquad (1.180)$$

CHAPTER 1. BASIC TECHNIQUES

$$C + E = -1 \tag{1.181}$$
$$\Rightarrow C = 0 \tag{1.182}$$

Combining this into the original integral gives

$$= \int \frac{1}{x} + \frac{x}{(x^2+1)^2} + \frac{-(x+1)}{x^2+1} \tag{1.183}$$

The first term is $\ln x$, the second one is a u-sub, and the third needs the negative distribution, the fraction split into two. The first of the resulting integrals is a u-sub, and the second one is a trig sub. Once you work through all of that, you arrive at the final answer,

$$\int \frac{1 - x + 2x^2 - x^3}{x(x^2+1)^2} \, dx = \ln x - \frac{1}{2(x^2+1)} - \tag{1.184}$$

$$\frac{1}{2} \ln(x^2+1) - \arctan x + C \tag{1.185}$$

Chapter 2

Substitutions

Now that you have a foundation (ideally strong) in the 4 major types of integration techniques that appear in numerous problems, we can start to dive into some more abstract, unintuitive substitutions. The part that makes them doable is that they are essentially just fancy u-subs. The changing of bounds, taking the derivative, etc., are all the same. The only difference is that you are not going to see the derivative of your substitution show up exactly in the actual integral (at least, it is very rare for this to happen). Some of these will be mind blowing in that they exist and work; others may seem more intuitive. Without further ado, let's examine some of these.

2.1 Weierstrauss Substitution

Karl Weierstrass was a German Mathematician who worked on a plethora of different problems and subjects. He is the one after whom this substitution is named. Additionally, he has come up with the Weierstrass function, which is continuous everywhere and differentiable nowhere. He also has a famous approximation theorem named after him.

The substitution is as follows.

> **Definition 2.1.1: W**
>
> ierstrass Substitution Let $f(x)$ be a function containing trigonometric functions.
>
> $$\int f(\sin x, \cos x)\, dx \qquad (2.1)$$
>
> Then, the following substitution can be made, Let
>
> $$t = \tan \frac{x}{2} \qquad (2.2)$$
>
> $$dt = \frac{1}{2}(1 + tan^2 \frac{x}{2})\, dx \qquad (2.3)$$
>
> $$dt = \frac{1}{2}(1 + t^2)\, dx \qquad (2.4)$$
>
> $$dx = \frac{2}{1 + t^2}\, dt \qquad (2.5)$$

The next goal is to derive the different trig functions in terms of t. Start with

$$\sin x = \frac{2 \sin \frac{x}{2} \cos \frac{x}{2}}{\cos^2 \frac{x}{2} + \sin^2 \frac{x}{2}} \qquad (2.6)$$

This is equal to
$$= \frac{2\tan\frac{x}{2}}{1+\tan^2\frac{x}{2}} \tag{2.7}$$

$$\Rightarrow \sin x = \frac{2t}{1+t^2} \tag{2.8}$$

For $\cos x$,
$$\cos x = \frac{\cos^2\frac{x}{2} - \sin^2\frac{x}{2}}{\cos^2\frac{x}{2} + \sin\frac{x}{2}} \tag{2.9}$$

$$= \frac{1-t^2}{1+t^2} \tag{2.10}$$

The other four trig functions can then be found by taking the reciprocals/ratios of these two.

Now, we can look at a few examples to show how useful this substitution is in practice.

Example 1

$$\int \csc x \, dx \tag{2.11}$$

Let

$$t = \tan \frac{x}{2} \tag{2.12}$$

$$dx = \frac{2}{1+t^2} \, dt \tag{2.13}$$

By flipping $\sin x$ as given above, it can be substituted into the integral.

$$\int \frac{1+t^2}{2t} \frac{2}{1+t^2} \, dt \tag{2.14}$$

This simplifies the integral to

$$\int \frac{1}{t} \, dt \tag{2.15}$$

Integrating and substituting t back in gives the final answer,

$$\int \csc x \, dx = \ln \tan \frac{x}{2} + C \tag{2.16}$$

This solution development was much easier than if you tried to do it the normal way. It would have required multiplying by ($\csc x - \cot x$ on the top and bottom and is very unintuitive.

Example 2

$$\int_0^{2\pi} \frac{1}{1+2\cos x} \, dx \tag{2.17}$$

Let

$$t = \tan \frac{x}{2} \tag{2.18}$$

$$dx = \frac{2}{1+t^2} dt \qquad (2.19)$$

Substituting this in along with cosine gives,

$$\int_{-\infty}^{\infty} \frac{2}{3+t^2} dt \qquad (2.20)$$

Then, make the substitution,

$$u = \sqrt{3}t \qquad (2.21)$$

$$du = \sqrt{3}\, dt \qquad (2.22)$$

This makes the integral,

$$\frac{2}{\sqrt{3}} \int_{-\infty}^{\infty} \frac{1}{1+u^2}\, du \qquad (2.23)$$

Therefore, the final value of the integral is

$$\int_0^{2\pi} \frac{1}{1+2\cos x}\, dx = \frac{2\pi}{\sqrt{3}} \qquad (2.24)$$

2.2 Euler Substitutions

Developed by Leonhard Euler, these substitutions help to solve integrals involving square roots of quadratic. They are very similar to trig subs in their scope. Personally, I think Euler substitutions are convoluted and are masked by how effective trig subs are. There are three main types of Euler substitutions that are used based on what exactly the quadratic looks like.

Definition 2.2.1: Euler's First Substitution

If the integral contains

$$\sqrt{ax^2 + bx + c} \tag{2.25}$$

and $a > 0$, then, let

$$\sqrt{ax^2 + bx + c} = x\sqrt{a} + t \tag{2.26}$$

Definition 2.2.2: Euler's Second Substitution

If $c > 0$, then let

$$\sqrt{ax^2 + bx + c} = xt + \sqrt{c} \tag{2.27}$$

Definition 2.2.3: Euler's Third Substitution

Given two roots of $ax^2 + bx + c$, α and β, you can substitute,
$$\sqrt{a(x-\alpha)(x-\beta)} = (x-\alpha)t \qquad (2.28)$$

However, the saving grace that makes an Euler Substitution viable by is allowing $a = 1$. Then, this simplifies it down to requiring only one substitution for any quadratic.
Let
$$x = \frac{c - t^2}{2t - b} \qquad (2.29)$$

Therefore,
$$dx = \frac{-2(t^2 - bt + c)}{(2t - b)^2} dt \qquad (2.30)$$

Therefore, this allows us to write,
$$\sqrt{x^2 + bx + c} = \frac{t^2 - bt + c}{2t - b} \qquad (2.31)$$

Let us look at a standard integral that would normally be solved using a trigonometric substitution and solve it using an Euler Substitution instead.

Example 2.2.1.

$$\int \sqrt{x^2+1}\, dx \tag{2.32}$$

For this example,

$$a = 1 \tag{2.33}$$
$$b = 0 \tag{2.34}$$
$$c = 1 \tag{2.35}$$

We can then substitute

$$x = \frac{1-t^2}{2t} \tag{2.36}$$

$$dx = \frac{-2(t^2+1)}{(2t)^2}\, dt \tag{2.37}$$

$$\sqrt{x^2+1} = \frac{t^2+1}{2t} \tag{2.38}$$

Now, we can substitute this back into the original integral.

$$\int \frac{t^2+1}{2t} \frac{-2(t^2+1)}{(2t)^2}\, dt \tag{2.39}$$

Simplifying,

$$= -\int \frac{(t^2+1)(t^2+1)}{4t^3}\, dt \tag{2.40}$$

$$= -\int \frac{t^4+2t^2+1}{4t^3}\, dt \tag{2.41}$$

$$= -\frac{1}{4}\int t + \frac{2}{t} + \frac{1}{t^3}\, dt \tag{2.42}$$

$$= -\frac{1}{4}(\frac{1}{2}t^2 + 2\ln|t| - \frac{1}{2t^2}) \qquad (2.43)$$

Substituting $t = \sqrt{x^2+1} - x$ back in gives us the final answer,

$$-\frac{1}{8}(\sqrt{x^2+1} - x)^2 - \frac{1}{2}\ln|\sqrt{x^2+1} - x| \qquad (2.44)$$

$$+\frac{1}{8(\sqrt{x^2+1} - x)^2} + C \qquad (2.45)$$

2.3 Hyperbolic Trig Substitution

This substitution is based entirely on the hyperbolic trigonometric functions. Essentially, instead of a unit circle, you have a unit hyperbola that defines the six functions. They have many of the same properties and identities as the regular trigonometric functions.

For example, the derivative of $\sinh x$ is $\cosh x$, and the derivative of $\cosh x$ is $\sinh x$. This makes it simpler than the regular trigonometric derivatives since you do not have to account for negative signs.

> **Definition 2.3.1: Hyperbolic Substitution**
>
> If an integral contains,
> $$\sqrt{x^2 - a^2} \qquad (2.46)$$
> Then, let
> $$x = a \cosh u. \qquad (2.47)$$
>
> If you have an integral that contains
> $$\sqrt{x^2 + a^2} \qquad (2.48)$$
> Then, let
> $$x = a \sinh x \qquad (2.49)$$

Then, you would utilize the hyperbolic trigonometric iden-

tities
$$\sinh^2 x - \cosh^2 x = 1 \qquad (2.50)$$
Also, you can write squares as
$$\sinh^2 x = \frac{1}{2}(-1 + \cosh 2x) \qquad (2.51)$$

$$\cosh^2 x = \frac{1}{2}(1 + \cosh 2x) \qquad (2.52)$$

$$\sinh 2x = 2 \sinh x \cosh x \qquad (2.53)$$

In general, this works very similarly to a regular trigonometric substitution.

Example 2.3.1.

$$\int \sqrt{x^2+1}\, dx \qquad (2.54)$$

Let

$$x = \sinh u \qquad (2.55)$$
$$dx = \cosh u\, du \qquad (2.56)$$

Substituting back in,

$$\int \cosh^2 u\, du \qquad (2.57)$$

Using the identity from above,

$$= \int \frac{1}{2}(1 + \cosh 2u)\, du \qquad (2.58)$$

Integrating gives,

$$\frac{1}{2}u + \frac{1}{4}\sinh 2u \qquad (2.59)$$

Then, using the double angle identity, and substituting in the inverse hyperbolic trig logarithm for $\sinh^{-1} x$, and $\cosh u = \sqrt{1+x^2}$, the final answer is,

$$\int \sqrt{x^2+1}\, dx = \frac{1}{2}\ln|x+\sqrt{x^2+1}| + \frac{1}{2}x\sqrt{x^2+1} + C \qquad (2.60)$$

CHAPTER 2. SUBSTITUTIONS

2.4 Ninja Substitution

In this section, we will discuss the ninja substitution. It is the first point in time when we see a substitution that is completely off-the-wall. I thought this was the neatest thing when I first learned about it. The process is as such,

---

Definition 2.4.1: Ninja Substitution

For an integral of the form,

$$\int_{-a}^{a} \frac{f(x)}{f(x) + g(x)} \, dx \qquad (2.61)$$

Where $f(-x) = g(x)$ and $g(-x) = f(x)$, let

$$u = -x \qquad (2.62)$$

This will still result in the exact same bounds, a trick that will be examined later on. Therefore, we find that

$$\int_{-a}^{a} \frac{f(x)}{f(x) + g(x)} \, dx = \int_{-a}^{a} \frac{g(x)}{g(x) + f(x)} \, dx \qquad (2.63)$$

Now, we add these two results, since the denominators are the exact same, we can just add the numerators.

$$\int_{-a}^{a} \frac{f(x) + g(x)}{f(x) + g(x)} \, dx \qquad (2.64)$$

$$= \frac{1}{2} \int_{-a}^{a} dx = a \qquad (2.65)$$

Generally speaking, the integral requires some further manipulation; however, there are exceptions in which everything will cancel out nicely.

Example 2.4.1.

$$\int_{-1}^{1} \frac{x^2}{1+e^x} \, dx \qquad (2.66)$$

Let

$$u = -x \qquad (2.67)$$

$$= \int_{-1}^{1} \frac{x^2}{1+e^{-x}} \, dx \qquad (2.68)$$

Multiply the top and bottom both by e^x,

$$= \int_{-1}^{1} \frac{x^2 e^x}{1+e^x} \, dx \qquad (2.69)$$

Now add this to the original integral,

$$2 \int_{-1}^{1} \frac{x^2}{1+e^x} \, dx = \int_{-1}^{1} \frac{x^2(1+e^x)}{1+e^x} \, dx \qquad (2.70)$$

Just like that, the terms cancel out.

$$\int_{-1}^{1} \frac{x^2}{1+e^x} \, dx = \frac{1}{2} \int_{-1}^{1} x^2 \qquad (2.71)$$

Therefore, the final answer is

$$\int_{-1}^{1} \frac{x^2}{1+e^x} \, dx = \frac{1}{3} \qquad (2.72)$$

CHAPTER 2. SUBSTITUTIONS

Example 2.4.2.

$$I = \int_0^\infty \frac{1}{(1+x^2)(1+e^x)} \, dx \qquad (2.73)$$

Substitute and rewrite,

$$= \int_0^\infty \frac{1}{(1+x^2)(1+e^{-x})} \, dx \qquad (2.74)$$

Multiply by e^x to the top and bottom and add the results.

$$2I = \int_0^\infty \frac{1+e^x}{(1+x^2)(1+e^x)} \, dx \qquad (2.75)$$

Canceling out the $1+e^x$ terms,

$$I = \frac{1}{2} \int_0^\infty \frac{1}{1+x^2} \, dx \qquad (2.76)$$

Therefore, the final answer is,

$$\int_0^\infty \frac{1}{(1+x^2)(1+e^x)} \, dx = \frac{\pi}{4} \qquad (2.77)$$

2.5 Inversion

Inversion is as equally interesting as the Ninja Substitution, if not more so. The statement is as such:

> **Definition 2.5.1: Inversion**
>
> For an integral,
> $$\int_0^\infty f(x)\,dx \tag{2.78}$$
> Make the substitution
> $$u = \frac{1}{x} \tag{2.79}$$
> $$du = -\frac{1}{x^2}\,dx \tag{2.80}$$

The bounds, once again, will be the same. Therefore, since the two integrals are equal, we can add them together to get,
$$2I = \int_0^\infty \frac{u^2 f(u) + f(\frac{1}{u})}{u^2}\,du \tag{2.81}$$
This applies especially well to natural logarithms. Since,
$$\ln\frac{1}{x} = -\ln x \tag{2.82}$$
Now, we can demonstrate this using two examples.

CHAPTER 2. SUBSTITUTIONS 51

Example 2.5.1.

$$\int_0^\infty \frac{\ln 2x}{1+x^2}\,dx \qquad (2.83)$$

Let

$$u = \frac{1}{x} \qquad (2.84)$$

$$du = -\frac{1}{x^2}dx \qquad (2.85)$$

Substitute this into the integral,

$$\int_0^\infty \frac{\ln \frac{2}{u}}{(1+\frac{1}{u^2})u^2}\,du \qquad (2.86)$$

Since u is just a dummy variable, meaning no matter what letter we subbed in, it means the exact same thing, we can safely change u back to x,

$$\int_0^\infty \frac{\ln \frac{2}{x}}{(1+\frac{1}{x^2})x^2}\,dx \qquad (2.87)$$

Now, add this integral and the original integral,

$$2I = \int_0^\infty \frac{\ln 2x + \ln \frac{2}{x}}{x^2+1}\,dx \qquad (2.88)$$

Dividing by 2 and simplifying gives,

$$I = \frac{1}{2}\int_0^\infty \frac{\ln 4}{x^2+1}\,dx \qquad (2.89)$$

Integrating gives,

$$I = \frac{\pi}{4}\ln 2 \qquad (2.90)$$

Rewriting using log rules gives the final answer,

$$\int_0^\infty \frac{\ln 2x}{1+x^2}\,dx = \frac{\pi}{2}\ln 2 \qquad (2.91)$$

Example 2.5.2.

$$\int_0^\infty \frac{\ln x}{x^2 + 2x + 4} \qquad (2.92)$$

Let

$$u = \frac{1}{x} \qquad (2.93)$$

$$du = -\frac{1}{u^2}\,dx \qquad (2.94)$$

Substituting back in,

$$\int_0^\infty \frac{\ln \frac{1}{x}}{(\frac{1}{x^2} + \frac{2}{x} + 4)x^2}\,dx \qquad (2.95)$$

There's an issue though, multiplying out the denominator does not give the same result as the denominator of the original integral. To fix this, substitute

$$u = \frac{c}{x} \qquad (2.96)$$

Then, after substituting this in, the integral becomes,

$$c\int_0^\infty \frac{\ln c - \ln x}{4x^2 + 2cx + c^2}\,dx \qquad (2.97)$$

Factor out a 4 in the denominator,

$$c\int_0^\infty \frac{\ln c - \ln x}{4(x^2 + \frac{2c}{4} + \frac{c^2}{4})}\,dx \qquad (2.98)$$

Now, set the x-coefficient to the x-coefficient in the original integral.

$$\frac{2c}{4} = 2 \qquad (2.99)$$

CHAPTER 2. SUBSTITUTIONS

$$\Rightarrow c = 4 \qquad (2.100)$$

Now, this value of c can be plugged into the integral, and, since the denominators are now the same, you can add the two integrals. Doing this gives,

$$2I = \int_0^\infty \frac{\ln 4}{x^2 + 2x + 4}\, dx \qquad (2.101)$$

To evaluate this integral, you can complete the square on the bottom and use a trig-sub.

$$I = \frac{1}{2}\int_0^\infty \frac{\ln 4}{(x+1)^2 + 3}\, dx \qquad (2.102)$$

The integral evaluates to

$$\ln 2 \left(\frac{1}{\sqrt{3}} \arctan\left(\frac{x+1}{\sqrt{3}}\right)\right)\Big|_0^\infty \qquad (2.103)$$

Plugging in the upper and lower bounds, we arrive at the final value of the integral,

$$\int_0^\infty \frac{\ln x}{x^2 + 2x + 4} = \frac{\pi \ln 2}{3\sqrt{3}} \qquad (2.104)$$

2.6 King's Rule

This section discusses another substitution that is very important for evaluating integrals. It is considered one of the most important tricks to have because it can make an absolute nightmare of an integral look much nicer. The statement is as follows,

> **Theorem 2.6.1: King's Rule**
>
> $$\int_a^b f(x)\, dx = \int_a^b f(a+b-x)\, dx \qquad (2.105)$$

We prove this by letting,

$$I = \int_a^b f(x)\, dx \qquad (2.106)$$

We can then substitute,

$$u = a + b - x \qquad (2.107)$$

$$du = -dx \qquad (2.108)$$

Substituting this back into the integral gives,

$$I = -\int_b^a f(u)\, du \qquad (2.109)$$

We can flip the bounds on the integral by absorbing the negative sign.

$$I = \int_a^b f(u)\, du \qquad (2.110)$$

Lastly, substituting in u we arrive at the conclusion,

$$\int_a^b f(x)\ dx = \int_a^b f(a+b-x)\ dx \qquad (2.111)$$

We can look at an example of how this is used in practice to simplify difficult integrals into more manageable ones.

Example 2.6.1.

$$\int_0^1 \frac{\ln(x+1)}{x^2+1} \tag{2.112}$$

First, let
$$x = \tan\theta \tag{2.113}$$
$$dx = 1 + \tan^2\theta \, d\theta \tag{2.114}$$
$$dx = 1 + x^2 \, d\theta \tag{2.115}$$

Substituting this in gives,

$$= \int_0^{\pi/4} \ln(1 + \tan\theta) \, d\theta \tag{2.116}$$

Now, apply King's Rule

$$u = \frac{\pi}{4} - x \tag{2.117}$$

Substituting this in and using the angle difference identity for tangent gives,

$$= \int_0^{\pi/4} \ln\left(\frac{\tan\frac{\pi}{4} - \tan\theta}{1 + \tan\frac{\pi}{4}\tan\theta} + 1\right) d\theta \tag{2.118}$$

Simplifying gives,

$$\int_0^{\pi/4} \ln\frac{2}{1+\tan\theta} \, d\theta \tag{2.119}$$

Now, split the logarithm using the division-subtraction rule.

$$\int_0^{\pi/4} \ln(1+\tan\theta) \, d\theta = \int_0^{\pi/4} \ln 2 \, d\theta - \int_0^{\pi/4} \ln(1+\tan\theta) \tag{2.120}$$

Since the integral on the left matches that of the second one on the right, we can add it over and divide the result by two. Lastly, integrate the first integral on the right-hand side to get the final result.

$$\int_0^1 \frac{ln(x+1)}{x^2+1} = \frac{\pi}{8} \ln 2 \qquad (2.121)$$

2.7 Queen's Rule

Queen's Rule is another substitution technique that is very similar to King's Rule. In fact, it is just a specific case of King's Rule. It involves the bounds of integration being 0 and $\frac{\pi}{2}$.

> **Theorem 2.7.1: Queen's Rule**
>
> $$\int_0^{\pi/2} f(x)\,dx = \int_0^{\pi/2} f(\frac{\pi}{2} - x)\,dx \qquad (2.122)$$

You can prove this by letting $a = 0$ and $b = \frac{\pi}{2}$ in King's Rule.

It is best applied with complex trigonometric functions and can often be paired very well with a Weierstrass substitution. Let's look at an example of it.

CHAPTER 2. SUBSTITUTIONS

Example 2.7.1.

$$I = \int_0^{\pi/2} \frac{\sin^3 x}{\sin^3 x + \cos^3 x} \, dx \qquad (2.123)$$

Apply Queen's Rule

$$I = \int_0^{\pi/2} \frac{\cos^3 x}{\sin^3 x + \cos^3 x} \, dx \qquad (2.124)$$

Now, we can add these together. Since the denominators are the same, we can just combine the numerators.

$$2I = \int_0^{\pi/2} \frac{\sin^3 x + \cos^3 x}{\sin^3 x + \cos^3 x} \, dx \qquad (2.125)$$

Just like that, the terms are completely canceled, leaving the final answer to the integral.

$$\int_0^{\pi/2} \frac{\sin^3 x}{\sin^3 x + \cos^3 x} \, dx = \frac{\pi}{4} \qquad (2.126)$$

Chapter 3

More Techniques

Now, we are truly growing in the types of problems we can solve. In this upcoming chapter, we will explore some additional general techniques that can be used to solve difficult integrals. Some of these have more use and value than others; however, they are still important to know in their own right. Some of these are so valuable to know that they can be used to solve almost any problem. Therefore, they are all good methods to have at your disposal so that you can undertake almost any integral. Ultimately, this chapter should help bridge the gap between elementary and advanced integration. A majority of the integrals will be solvable using either one of the methods in this chapter or, more simply, using one of our previous methods. Using different methods may open up new important results that can be applied to many different problems. Without any further ado, let's look at some abstract integration techniques.

3.1 Feynman's Technique

This method is somewhat of a misnomer, as Richard Feynman, an American physicist, was not technically the first to pioneer it, though he did swear by it as a remarkable integration technique. It is also known by its true name, the Leibniz Integral Rule. He was one of the major co-founders of calculus, as we know it, alongside Isaac Newton. The theorem behind it is very inconspicuous, and you may question how exactly it works.

> **Theorem 3.1.1: Leibniz Integral Rule**
>
> Let $f(x,t)$ be a function such that both it and its partial derivative $f_x(x,t)$ are continuous in the xt-plane. Then,
>
> $$\frac{d}{dx}(\int_{a(x)}^{b(x)} f(x,t) \ dt) = f(x,b(x))b'(x)- \quad (3.1)$$
>
> $$f(x,a(x))a'(x) + \int_{a(x)}^{b(x)} f_x(x,t) \ dt \quad (3.2)$$

Now, I can almost hear what you are thinking. What is this humble-jumble mess of symbols, and better yet, why is this important? Well, in the limiting case that you assume $a(x) = a$ and $b(x) = b$ for all x, this drastically simplifies

CHAPTER 3. MORE TECHNIQUES

the statement down to,

$$\frac{d}{dx}(\int_a^b f(x,t)\ dt) = \int_a^b f_x(x,t)\ dt \qquad (3.3)$$

This the the Feynman Trick that we utilize for solving integrals.

In order to prove this theorem, we start with the statement

$$\int_x^{x+h} \int_a^b f_x(x,t)\ dt\ dx \qquad (3.4)$$

Fubini's Theorem states that, as long as these functions are convergent over the bounds of integration and the integral converges, you can change the order in which you perform the double integration. Using this theorem, we find that

$$\int_x^{x+h} \int_a^b f_x(x,t)\ dt\ dx = \int_a^b \int_x^{x+h} f_x(x,t)\ dx\ dt \qquad (3.5)$$

Using the Fundamental Theorem of Calculus, we find that

$$\int_x^{x+h} \int_a^b f_x(x,t)\ dt\ dx = \int_a^b f(x+h,t)\ dt - \int_a^b f(x,t)\ dt \qquad (3.6)$$

You may recognize that if we divide by h, and take the limit as $h \to 0$, we obtain the derivative. Doing this gives,

$$\frac{\int_a^b f(x+h,t)\ dt - \int_a^b f(x,t)\ dt}{h} = \frac{1}{h} \int_a^b \int_x^{x+h} f_x(x,t)\ dx\ dt \qquad (3.7)$$

Define
$$F(u) := \int_{x_0}^{u} \int_{a}^{b} f_x(x,t) \, dt \, dx \qquad (3.8)$$

Therefore, everything can be rewritten as

$$\frac{\int_{a}^{b} f(x+h,t) \, dt - \int_{a}^{b} f(x,t) \, dt}{h} = \frac{F(x+h) - F(x)}{h} \qquad (3.9)$$

Now, take the limit as $h \to 0$ on both sides. The left side becomes,

$$\frac{d}{dx}\left(\int_{a}^{b} f(x,t) \, dt\right) \qquad (3.10)$$

The right side then becomes,

$$\int_{a}^{b} f_x(x,t) \, dt \qquad (3.11)$$

Now, setting them equal, we have arrived at our expected result

$$\frac{d}{dx}\left(\int_{a}^{b} f(x,t) \, dt\right) = \int_{a}^{b} f_x(x,t) \, dt \qquad (3.12)$$

Essentially, it states that if you define an extra parameter in your integral, you can take the derivative with respect to it, integrate it, and it will yield the exact same result. Then, you would have the derivative of your integral, so you can take the indefinite integral with respect to that parameter and plug in whatever value you need.

CHAPTER 3. MORE TECHNIQUES

Example 3.1.1.

Evaluate
$$\int_{-\pi/2}^{\pi/2} \frac{\ln(1 - a\sin x)}{\sin x} \, dx \qquad (3.13)$$

where a is between -1 and 1.

First, we let
$$I(a) = \int_{-\pi/2}^{\pi/2} \frac{\ln(1 - a\sin x)}{\sin x} \, dx \qquad (3.14)$$

Now, we apply the formula that we derived above. This allows us to find $I'(a)$.

$$I'(a) = -\int_{-\pi/2}^{\pi/2} \frac{1}{1 - a\sin x} \, dx \qquad (3.15)$$

Next, use a Weierstrauss Substitution.

$$I'(a) = -\int_{0}^{1} \frac{1}{1 - a\frac{2t}{1+t^2}} \frac{2}{1+t^2} \, dt \qquad (3.16)$$

$$= -\int_{0}^{1} \frac{2}{1 + t^2 - a(2t)} \, dt \qquad (3.17)$$

$$= -2\int_{0}^{1} \frac{1}{(t-a)^2 + (1-a)^2} \, dt \qquad (3.18)$$

This requires using a trigonometric substitution. Doing the work for this gives,

$$I'(a) = \frac{-2}{\sqrt{1-a^2}}\left(\arctan\sqrt{\frac{1+a}{1-a}} + \arctan\sqrt{\frac{1+a}{1-a}}\right) \qquad (3.19)$$

Now, using the identity,

$$\arctan \frac{1}{x} + \arctan x = \frac{\pi}{2} \tag{3.20}$$

Therefore, this gives the equation for $I'(a)$.

$$I'(a) = \frac{-\pi}{\sqrt{1-a^2}} \tag{3.21}$$

Now, we can integrate with respect to a on both sides to get our original integral in terms of a.

$$\int I'(a)\,da = \int \frac{-\pi}{\sqrt{1-a^2}}\,da \tag{3.22}$$

This gives the equation for $I(a)$,

$$I(a) = -\pi \arcsin a + C \tag{3.23}$$

In order to find C, the constant of integration, refer back to the integral definition of $I(a)$.

$$\int_{-\pi/2}^{\pi/2} \frac{\ln(1-a\sin x)}{\sin x}\,dx \tag{3.24}$$

If we plug in $a = 0$ it becomes $\ln 1$ in the numerator, causing the entire integral to collapse to 0. Therefore,

$$I(0) = 0 = -\pi \arctan(0) + C \tag{3.25}$$

$$\Rightarrow C = 0 \tag{3.26}$$

Now, we can write the final expression for the integral only in terms of a.

$$\int_{-\pi/2}^{\pi/2} \frac{\ln(1-a\sin x)}{\sin x}\,dx = -\pi \arcsin a \tag{3.27}$$

where a is between -1 and 1.

CHAPTER 3. MORE TECHNIQUES

Example 3.1.2.

$$\int_{-1}^{1} \arctan x \arcsin x \, dx \tag{3.28}$$

First, since this is an even function, we can rewrite the symmetric bounds.

$$2\int_{0}^{1} \arctan x \arcsin x \, dx \tag{3.29}$$

Next, make the substitution

$$u = \arcsin x \tag{3.30}$$

$$x = \sin u \tag{3.31}$$

$$dx = \cos u \, du \tag{3.32}$$

Now, substitute this back into the integral.

$$I = 2\int_{0}^{\pi/2} u \arctan(\sin u) \cos u \, du \tag{3.33}$$

Now, use integration by parts, giving the result,

$$I = 2u \sin u \arctan \sin u \big|_{0}^{\pi/2} \tag{3.34}$$

$$-2\int_{0}^{\pi/2} \sin u \arctan \sin u + \frac{u \cos u}{1 + \sin^2 u} \sin u \, du \tag{3.35}$$

Then, after plugging in the bounds of the first term and separating the integral into two separate ones, you get,

$$I = \frac{\pi^2}{4} - 2\int_{0}^{\pi/2} \sin u \arctan \sin u \, du - \tag{3.36}$$

$$2\int_0^{\pi/2} \frac{u\cos u}{1+\sin^2 u} \sin u\, du \qquad (3.37)$$

Let

$$I_1 = \int_0^{\pi/2} \sin u \arctan \sin u\, du \qquad (3.38)$$

Let

$$I_2 = \int_0^{\pi/2} \frac{u\cos u}{1+\sin^2 u} \sin u\, du \qquad (3.39)$$

Now, lets work on evaluating each of these two integrals, starting with I_1. First, we can start by using integration by parts on it.

$$I_1 = -\arctan(\sin u)\cos u\Big|_0^{\pi/2} + \int_0^{\pi/2} \frac{\cos^2 u}{1+\sin^2 u}\, du \qquad (3.40)$$

The second integral could then be solved by a Weierstrauss Substitution. Doing this and evaluating the bounds of the first term will give you the final evaluation for I_1.

$$I_1 = \frac{\pi}{2}(\sqrt{2} - 1) \qquad (3.41)$$

Now that we have evaluated the first integral, we can work on the second one. First, use integration by parts.

$$I_2 = \frac{1}{2}u\ln(1+\sin^2 u)\Big|_0^{\pi/2} - \frac{1}{2}\int_0^{\pi/2} \ln(1+\sin^2 u)\, du \qquad (3.42)$$

Therefore, after evaluating the first time, we arrive at

$$\frac{\pi}{4}\ln 2 - \frac{1}{2}\int_0^{\pi/2} \ln(1+\sin^2 u)\, du \qquad (3.43)$$

CHAPTER 3. MORE TECHNIQUES

Now, make the substitution,

$$z = \cot u \qquad (3.44)$$

$$dz = \csc^2 u \, du \qquad (3.45)$$

Substitute this back into I_2.

$$I_2 = \frac{\pi}{4} \ln 2 + \frac{1}{2} \int_0^\infty \frac{\ln(1 + \frac{1}{1+z^2})}{1 + z^2} \, dz \qquad (3.46)$$

$$= \frac{\pi}{4} \ln 2 + \frac{1}{2} \int_0^\infty \frac{\ln(1 + z^2)}{1 + z^2} \, dz - \frac{1}{2} \int_0^\infty \frac{\ln(2 + z^2)}{1 + z^2} \, dz \qquad (3.47)$$

We have two more gross looking integrals that we have to evaluate. Let

$$I_3 = \frac{1}{2} \int_0^\infty \frac{\ln(1 + z^2)}{1 + z^2} \, dz \qquad (3.48)$$

Let

$$I_4 = \frac{1}{2} \int_0^\infty \frac{\ln(2 + z^2)}{1 + z^2} \, dz \qquad (3.49)$$

Start with I_3 by letting

$$w = \tan z \qquad (3.50)$$

Substituting this in gives,

$$I_3 = -2 \int_0^{\pi/2} \ln \cos w \, dw \qquad (3.51)$$

This is another famous integral that evaluates to $-\frac{\pi}{2} \ln 2$. Therefore,

$$I_3 = \pi \ln 2 \qquad (3.52)$$

Moving on, we can look at I_4, where we can finally utilize Feynman's technique. First, define $I(a)$.

$$I(a) = \int_0^\infty \frac{\ln(1+a^2x^2)}{1+x^2}\, dx \qquad (3.53)$$

Now, apply the theorem from above.

$$I'(a) = \int_0^\infty \frac{2ax^2}{(1+x^2)(1+a^2x^2)}\, dx \qquad (3.54)$$

Evaluating this integral gives

$$I'(a) = \left(\frac{2a \arctan x}{a^2-1} - \frac{a \arctan ax}{a^2-1}\right)\Big|_0^\infty = \frac{\pi}{1+a} \qquad (3.55)$$

$$I'(a) = \frac{\pi}{1+a} \qquad (3.56)$$

Now, integrate this on both sides with respect to a.

$$I(a) = \pi \ln(1+a) + C \qquad (3.57)$$

Solving for C, just as in the last example, we find that

$$C = 0 \qquad (3.58)$$

$$I_4 = \ln 2 \int_0^\infty \frac{1}{1+z^2}\, dz + I\left(\sqrt{\frac{1}{2}}\right) \qquad (3.59)$$

$$I_4 = \frac{\pi \ln 2}{2} + \pi \ln\left(1 + \frac{1}{\sqrt{2}}\right) \qquad (3.60)$$

Combining this result with I_3, we can find the value of I_2.

$$I_2 = \frac{\pi \ln 2}{2} - \frac{\pi \ln(1+\frac{1}{\sqrt{2}})}{2} \qquad (3.61)$$

CHAPTER 3. MORE TECHNIQUES

Therefore, we can finally combine the results of all our work.

$$\int_{-1}^{1} \arctan x \arcsin x \, dx = \frac{\pi^2}{4} - \pi(\sqrt{2} - 1) \qquad (3.62)$$

$$-\frac{3\pi \ln 2}{2} + \pi \ln(1 + \sqrt{2}) \qquad (3.63)$$

3.2 "Sum" Crazy Technique

Now that we have discovered Feynman's Trick, we can explore some different tactics for solving an equal number of integrals. In this section, we will look at how infinite sums can be used to solve difficult integrals. The hardest part of evaluating some of these is actually knowing how to evaluate the sum, not the integral. A lot of the values will be defined by special functions and/or famous derivations. It comes down to recognizing where the sum originates.

Before doing integrals, we need to look at what these series are, when we can and cannot use them, and exactly how to use them. In general terms, we will be talking about the Maclaurin Series of a function. In simple terms, they are an infinite sum of polynomials that, when added together, equals a function. They are found by taking consecutive derivatives of the function, dividing them by a specific number, and adding them together.

$$f(x) = f(0) + f'(0) + \frac{f''(0)}{2!} + \frac{f^{(3)}(0)}{3!} + ... \qquad (3.64)$$

For example, take $f(x) = e^x$.

$$e^x = \sum_{n=0}^{\infty} \frac{x^n}{n!} \qquad (3.65)$$

We can also compose this function, for example,

$$e^{-x} = \sum_{n=0}^{\infty} (-1)^n \frac{x^n}{n!} \qquad (3.66)$$

CHAPTER 3. MORE TECHNIQUES

Or even,

$$e^{-x^2} = \sum_{n=0}^{\infty} (-1)^n \frac{x^{2n}}{n!} \tag{3.67}$$

This generally only applies if the function in question is continuous. Also, if you switch the order of the integral and the sum, it must abide by Fubini's Theorem (which requires continuity and convergence over the interval on which it is integrated). For example, if the series expansion of $f(x)$ is only defined over $[-1, 1]$, then you cannot integrate over $(-\infty, \infty)$.

This also applies to numerous other functions, which allows this technique of integration to be so useful. To make life simpler, as special functions arise in evaluating the sums in the integrals, I will define them and provide a brief excerpt about them. The goal of this is to make learning about their uses in integration much easier later on.

Example 3.2.1.

$$\int_0^1 \ln x \arctan x \, dx \qquad (3.68)$$

Substitute the Maclaurin series for $\arctan x$.

$$= \int_0^1 \ln x \sum_{k=0}^{\infty} (-1)^k \frac{x^{2k+1}}{2k+1} \, dx \qquad (3.69)$$

Now, since this cooperates with Fubini's Theorem, we can change the order of the sum and integral.

$$I = \int_0^1 \sum_{k=0}^{\infty} \ln x (-1)^k \frac{x^{2k+1}}{2k+1} \, dx \qquad (3.70)$$

$$I = \sum_{k=0}^{\infty} \frac{(-1)^k}{2k+1} \int_0^1 x^{2k+1} \ln x \, dx \qquad (3.71)$$

To solve the integral, use integration by parts.

$$= \frac{x^{2k+1}}{2k+1} \ln x \Big|_0^1 - \frac{1}{2k+2} \int_0^1 x^{2k+1} \, dx \qquad (3.72)$$

The first term evaluates to 0, and the second evaluates to

$$= \frac{-1}{(2k+2)^2} \qquad (3.73)$$

Now, the problem that was originally an integral has become a sum. Define it by S.

$$S = \sum_{k=0}^{\infty} \frac{(-1)^k}{(2k+1)(2k+2)^2} \qquad (3.74)$$

CHAPTER 3. MORE TECHNIQUES

First, we can split this up by using a partial fraction decomposition. This gives,

$$I = -\sum_{k=0}^{\infty} \frac{(-1)^k}{2k+1} + \frac{1}{2}\sum_{k=0}^{\infty} \frac{(-1)^k}{k+1} + \frac{1}{4}\sum_{k=0}^{\infty} \frac{(-1)^k}{(k+1)^2} \quad (3.75)$$

The first sum is $\arctan x$ evaluated at $x = 1$. The second sum evaluates to $\ln 2$, and the third sum evaluates to $\frac{\pi^2}{12}$. Combining these results gives the final answer.

$$\int_0^1 \ln x \arctan x \, dx = \frac{1}{2}\ln 2 + \frac{\pi^2}{48} - \frac{\pi}{4} \quad (3.76)$$

Example 3.2.2.

$$I = \int_0^{\pi/2} \ln \cos x \, dx \qquad (3.77)$$

Consider the sum,

$$\ln 2 \cos x = \sum_{n=1}^{\infty} \frac{(-1)^{n+1} \cos(2nx)}{n} \qquad (3.78)$$

We can solve for $\ln \cos x$.

$$\ln \cos x = -\ln 2 + \sum_{n=1}^{\infty} \frac{(-1)^{n+1} \cos(2nx)}{n} \qquad (3.79)$$

Now, we can substitute this into the integral.

$$I = \int_0^{\pi/2} -\ln 2 \, dx + \sum_{n=1}^{\infty} \frac{(-1)^{n+1}}{n} \int_0^{\pi/2} \cos(2nx) \, dx \quad (3.80)$$

Now, you can look at the graph of $\cos 2nx$, from that, you can see that the area underneath of the x-axis is equal to the area above it for any value of n. Therefore, the integral above collapses to zero, simplifying our solution to our final answer.

$$\int_0^{\pi/2} \ln \cos x \, dx = -\frac{\pi}{2} \ln 2 \qquad (3.81)$$

You can also show that

$$\int_0^{\pi/2} \ln \sin x \, dx = -\frac{\pi}{2} \ln 2 \qquad (3.82)$$

by using Queen's Rule.

CHAPTER 3. MORE TECHNIQUES

Additionally, using the properties of logarithms, it follows that
$$\int_0^{\pi/2} \ln \tan x \, dx = 0 \tag{3.83}$$
and
$$\int_0^{\pi/2} \ln \cot x \, dx = 0 \tag{3.84}$$

3.3 Ramanujan's Master Theorem

Ramanujan's Master Theorem, named after Srinivasa Ramanujan, is a method that finds the Mellin Transform of a function.

> **Theorem 3.3.1: Ramanujan Master Theorem**
>
> he Mellin Transform of a function $f(x)$ is given by
>
> $$\int_0^\infty x^{s-1} f(x) \, dx = \Gamma(s)\phi(-s) \qquad (3.85)$$

As we will find in a later section, Γ is known as the Gamma function. In short, it evaluates the factorial of a function. The only difference, though, is that it can be expanded to work for more than just positive integers. You can even find $\frac{1}{2}!$.

The second term, $\phi(-s)$, is used in a series expansion to determine the coefficients. As shown in the equation below, it is part of an infinite series.

$$f(x) = \sum_{k=0}^{\infty} \frac{\phi(k)}{k!} (-x)^k \qquad (3.86)$$

The ultimate strategy will be to determine what $\phi(k)$ is. Generally, you can examine this by looking at the series expansion of $f(x)$ and finding a pattern. This could involve alternating signs, specific patterns of coefficients, etc. Then,

CHAPTER 3. MORE TECHNIQUES

you can write this in terms of k, not x, and use this to evaluate the integral. Consider the following proof of this theorem.

First, we have to get some vocabulary and definitions out of the gate. Define E to be the "forward-shift operator". It simply advances the function a step forward.

$$E\phi(k) = \phi(k+1) \tag{3.87}$$

Rewriting $f(x)$ as the series expansion given above,

$$\int_0^\infty x^{s-1} \sum_{k=0}^\infty \frac{\phi(k)}{k!}(-x)^k \, dx \tag{3.88}$$

Then, using the forward-shift operator, this is equal to

$$\phi(0) \int_0^\infty x^{s-1} \sum_{k=0}^\infty \frac{(-1)^k}{k!} E^k x^k \, dx \tag{3.89}$$

The sum conveniently simplifies since it is the Maclaurin Series of e^{-Ex}. Therefore, this simplifies to

$$\int_0^\infty x^{s-1} e^{-Ex} \, dx \; \phi(0) \tag{3.90}$$

Then, using the definition of the Gamma Function (which we will talk about later), we can get an expression in terms of $\Gamma(s)$.

$$\Gamma(s) = \int_0^\infty x^{s-1} e^{-x} \, dx \tag{3.91}$$

This turns the integral into

$$\frac{\Gamma(s)}{E^s}\phi(0) \qquad (3.92)$$

Dividing by the forward-shift operator s times turns $\phi(0) \to \phi(-s)$. Therefore, this gives us the final form

$$\int_0^\infty x^{s-1} f(x)\, dx = \Gamma(s)\phi(-s) \qquad (3.93)$$

ending our proof with Ramanujan's Master Theorem.

Now that we have the "why" behind the theorem, we can see how it can actually be applied to specific, and generalized integrals in a fascinating way.

CHAPTER 3. MORE TECHNIQUES

Example 3.3.1.

$$\int_0^\infty x^{s-1} \sin x \, dx \tag{3.94}$$

So, we know what our $f(x)$ is, so we can set the maclaurin series of $\sin x$ off to the side to compare it to what we need for $\phi(k)$.

$$\sin x = \sum_{k=0}^\infty (-1)^k \frac{x^{2k+1}}{(2k+1)!} \tag{3.95}$$

$$f(x) = \sum_{k=0}^\infty \frac{\phi(k)}{k!}(-x)^k \tag{3.96}$$

Now, expand out $\sin x$ term by term.

$$\sin x = 0 + x + 0x^2 - x^3 + 0x^4 + \ldots \tag{3.97}$$

I am omitting the factorials underneath for simplicities sake. As you can see, there is a pattern with the signs that emerge. The values of the coefficients follow the pattern

$$0, 1, 0, -1, 0, 1\ldots \tag{3.98}$$

This, then, can be encapsulated by taking an oscillating function (such as $\sin x$), and multiplying the coefficient inside the argument to get 0, 1, or -1 for $x = 1, 2, 3\ldots$. The function in question?

$$\phi(-k) = \sin(\frac{\pi}{2}k) \tag{3.99}$$

Now that we have found our function, we can plug this into our formula given by Ramanujan's Master Theorem. We then find that the integral evaluates to

$$\int_0^\infty x^{s-1} \sin x \, dx = \Gamma(s) \sin(\frac{\pi}{2}s) \qquad (3.100)$$

In the end, the sine term is only truly a placeholder. It doesn't affect the actual value that the integral takes for different values of s, rather, it only determines the sign.

3.4 Frullani's Integrals

First developed by the Italian mathematician G. Frullani, this "theorem" strongly generalizes many different integrals. It can be used to help determine the value of some challenging integrals just by knowing this trick. The statement is as follows,

> **Theorem 3.4.1: Frullani's Theorem**
>
> For $f(x)$, a continuous, differentiable function over $[0, \infty)$, the following integral relation applies.
>
> $$\int_0^\infty \frac{f(ax) - f(bx)}{x} \, dx = (f(\infty) - f(0)) \ln(\frac{a}{b}) \quad (3.101)$$

This allows us to find the value of an integral that may not be solvable using any techniques by employing this formula. We can even find the integral if we don't know what the function is, or if we don't exactly know what it is.

Introducing a parameter, t, we can rewrite the integrand as a double integral rather than a single one. By using,

$$\frac{\partial}{\partial t}(\frac{f(xt)}{x}) \quad (3.102)$$

$$\frac{f(ax) - f(bx)}{x} = \frac{f(xt)}{x}\Big|_a^b \quad (3.103)$$

Writing this as an integral,

$$\frac{f(ax) - f(bx)}{x} = \int_a^b f'(xt) \, dt \quad (3.104)$$

3.4. FRULLANI'S INTEGRALS

Now, integrate on both sides with respect to x.

$$\int_0^\infty \frac{f(ax) - f(bx)}{x} \, dx = \int_0^\infty \int_a^b f'(xt) \, dt \, dx \quad (3.105)$$

Now, since the function is generally well behaved over the interval we are integrating on, we can change the order of integration.

$$\int_0^\infty \frac{f(ax) - f(bx)}{x} \, dx = \int_a^b \int_0^\infty f'(xt) \, dx \, dt \quad (3.106)$$

Then, we can integrate the right hand side with respect to x. This allows us to find an expression in terms of t, which we will then be able to integrate with respect to t.

$$\int_0^\infty \frac{f(ax) - f(bx)}{x} \, dx = \int_a^b \frac{f(xt)}{t} \Big|_0^\infty \, dt \quad (3.107)$$

$$\int_0^\infty \frac{f(ax) - f(bx)}{x} \, dx = \int_a^b \frac{f(\infty) - f(0)}{t} \, dt \quad (3.108)$$

Lastly, integrating this with respect to t gives us the final form of our Frullani Integral.

$$\int_0^\infty \frac{f(ax) - f(bx)}{x} \, dx = (f(\infty) - f(0)) \ln\left(\frac{a}{b}\right) \quad (3.109)$$

Now that we know it works, we can see exactly where and how it is useful. First, you can derive an integral representation of the natural logarithm by letting $f(x) = e^{-x}$ and $a = 1$.

$$\int_0^\infty \frac{e^{-x} - e^{-bx}}{x} \, dx \quad (3.110)$$

Using Frullani's formula,

$$\int_0^\infty \frac{e^{-x} - e^{-bx}}{x}\,dx = (e^{-\infty} - e^0)\ln\left(\frac{1}{b}\right) \qquad (3.111)$$

Since $e^{\infty} = 0$ and $\ln(\frac{1}{b}) = -\ln b$, we arrive at the integral representation of the natural logarithm of b.

$$\int_0^\infty \frac{e^{-x} - e^{-bx}}{x}\,dx = \ln b \qquad (3.112)$$

Let's try another example in which the original integrand does not appear to fall under the blanket of the "Frullani Integral".

Example 3.4.1.

$$\int_0^\infty \frac{\sin(px)\sin(qx)}{x}\,dx \qquad (3.113)$$

To get this into a form that matches Frullani's, utilize the identity:

$$\sin(px)\sin(qx) = \frac{1}{2}(\cos(p-q)x - \cos(p+q)x) \qquad (3.114)$$

Now, substituting this in, we see that it matches the Frullani Integral much nicer.

$$I = \frac{1}{2}\int_0^\infty \frac{\cos(p-q)x - \cos(p+q)x}{x}\,dx \qquad (3.115)$$

There is an issue, though. We can't directly apply the formula becuase $\cos(\infty)$ does not exist. To counteract this, introduce an additional term onto $\cos x$. Let

$$f(x) = e^{-xt}\cos x \qquad (3.116)$$

Now, if we let $t \to 0$, this still gives us exactly what we are looking for, plus avoiding the issue with $\cos\infty$. Combining this with the formula, we arrive at the final answer for the generalized integral.

$$\int_0^\infty \frac{\sin(px)\sin(qx)}{x}\,dx = \ln\frac{p+q}{p-q} \qquad (3.117)$$

Additionally, there is another extension/simplification of this theorem if $f(\infty)$ approaches 0.

$$\int_0^\infty \frac{f(ax) - f(bx)}{x}\,dx = f(0)\ln(\frac{b}{a}) \qquad (3.118)$$

CHAPTER 3. MORE TECHNIQUES

Start with the statement proved above.

$$\int_0^\infty \frac{f(ax) - f(bx)}{x} \, dx = (f(\infty) - f(0)) \ln\left(\frac{a}{b}\right) \quad (3.119)$$

Since we are stating that the $f(\infty)$ term goes to zero, we can eliminate it, simplifying the formula to

$$\int_0^\infty \frac{f(ax) - f(bx)}{x} \, dx = -f(0) \ln\left(\frac{a}{b}\right) \quad (3.120)$$

Now, using the logarithm rules, we can rewrite this in its final simplified form.

$$\int_0^\infty \frac{f(ax) - f(bx)}{x} \, dx = f(0) \ln\left(\frac{b}{a}\right) \quad (3.121)$$

This would best apply to the integral representation of the natural logarithm, as shown above. Using this formula is not a requirement; it is only a simplification. Up above, we were able to solve the exact same problem without the simplified formula, demonstrating that it is functionally the same.

3.5 Lobachevsky's Rule

This trick is a generalization of the Dirichlet Integral

$$\int_0^\infty \frac{\sin x}{x}\, dx = \frac{\pi}{2} \qquad (3.122)$$

> **Theorem 3.5.1: Lobachevsky's Rule**
>
> If $f(x)$ is an even, π-periodic function, then
>
> $$\int_0^\infty \frac{\sin x}{x} f(x)\, dx = \int_0^{\pi/2} f(x)\, dx \qquad (3.123)$$

Additionally, Lobachevsky found that

$$\int_0^\infty \frac{\sin^2 x}{x^2} f(x)\, dx = \int_0^{\pi/2} f(x)\, dx \qquad (3.124)$$

The proof of these two rules involves Fourier Series. I won't prove it in this book, however, but the proof is quite interesting to read about.

Example 3.5.1.

$$\int_0^{\pi/2} \frac{\sin \tan x}{\tan x} \, dx \qquad (3.125)$$

First, make the substitution,

$$u = \tan x \qquad (3.126)$$

$$du = \sec^2 u \, dx \qquad (3.127)$$

Substituting this back in,

$$I = \int_0^\infty \frac{\sin u}{u} \cos^2 u \, du \qquad (3.128)$$

Now, utilize Lobachevsky's Rule.

$$I = \int_0^\infty \frac{\sin u}{u} \cos^2 u \, du = \int_0^{\pi/2} \cos^2 u \, du \qquad (3.129)$$

Great! We only have to integrate the right-hand side. Doing this gives us

$$I = \frac{u}{2} + \frac{\sin 2u}{4} \Big|_0^{\pi/2} \qquad (3.130)$$

Substituting in the bounds, we get the final answer to the integral.

$$\int_0^{\pi/2} \frac{\sin \tan x}{\tan x} \, dx = \frac{\pi}{4} \qquad (3.131)$$

3.6 Polar Coordinates

Suppose you have an integral that is completely unsolvable using any of the previous techniques. There is one more technique that you can try to solve the integral. Or, suppose you have a double integral, you can try using this method in order to evaluate it.

> **Definition 3.6.1: Polar Substitution**
>
> If you have a function, $f(x,y)$, you can make the substitutions
> $$x = r\cos\theta \qquad (3.132)$$
> $$y = r\sin\theta \qquad (3.133)$$
> Then, the integral will change form.
> $$\iint_{R_1} f(x,y)\ dx\ dy = \iint_{R_2} f(r\cos\theta, r\sin\theta) r\ dr\ d\theta \qquad (3.134)$$

Much like what we have seen in all of the substitutions, the bounds change when we make substitutions. This still applies for multiple integrals.

This is especially useful when you are integrating over regions that are circles or if $f(x,y)$ takes the form
$$f(x,y) = x^2 + y^2 \qquad (3.135)$$
Then, you can show that
$$r^2 = x^2 + y^2 \qquad (3.136)$$

Since,
$$x^2 + y^2 = r^2 \cos^2 \theta + r^2 \sin^2 \theta \qquad (3.137)$$
Using the Pythagorean Identity,
$$r^2 = x^2 + y^2 \qquad (3.138)$$

Example 3.6.1.

$$\iint_R \sin(x^2 + y^2) \, dx \, dy \tag{3.139}$$

where $R : 1 \leq x^2 + y^2 \leq 16$. Now, we can make the substitution from above. Doing this transforms the integral to,

$$\iint r \sin(r^2) \, dr \, d\theta \tag{3.140}$$

Using the fact that R was a circle, we know that θ ranges from $[0, 2\pi]$, and if $1 \leq R^2 \leq 16$, then $1 \leq R \leq 4$. Therefore, the bounds of the integral become,

$$\int_0^{2\pi} \int_1^4 r \sin r^2 \, dr \, d\theta \tag{3.141}$$

Now, using a u-substitution, the first integral evaluates to

$$I = \int_0^{2\pi} (-\cos 16 + \cos 1) \, d\theta \tag{3.142}$$

Lastly, we can simply integrate both terms since they are both constants. This gives us our final answer to the integral.

$$\iint_R \sin(x^2 + y^2) \, dx \, dy = 2\pi \cos 1 - 2\pi \cos 16 \tag{3.143}$$

Chapter 4

Special Functions

We have seen some interesting techniques in the last chapter, and introduced one of our special functions. There is much more that can be done with them. Personally, combined with Feynman's Technique, special functions are one of the most useful methods of integrating. There are 4 main ones that we will look at: Gamma, Beta, Polygamma, and Zeta. We will look at each of them one-by-one and work through an example of them. The possibilities are almost endless as to what you can do with them, so this book will only scratch the surface. If you find these functions interesting, I highly recommend reading more about them. The Zeta Function, in particular is fascinating because of the unsolved "Riemann Hypothesis" regarding its non-trivial zeros. Either way, let's get into the definitions and properties of some of these special functions that allow us to use them for solving difficult integrals.

4.1 Gamma Function

The Gamma Function was first coined and defined by Daniel Bernoulli. It is an analytic continuation of the factorial function to all complex numbers (with the exception of negative integers). In short, it is defined by,

> **Definition 4.1.1: Factorial Definition**
>
> $$\Gamma(n) = (n-1)! \tag{4.1}$$

Why he chose $(n-1)!$ and not $n!$ is a mystery. Personally, I think he did it to mess with people.

> **Definition 4.1.2: Integral Definition**
>
> The Gamma Function is defined by an improper integral
>
> $$\Gamma(z) = \int_0^\infty t^{z-1} e^{-t} \, dt \tag{4.2}$$

You may be thinking (or not) that this looks very similar to the Mellin Transform seen back when talking about Ramanujan's Master Theorem. You would be correct! In fact, $\Gamma(z)$ is just the Mellin Transform of e^{-x}.

Another formulation of the function arises from the functional equation. As you may know, the factorial of $n-1$ is just

$$(n-1)! = (n-1)(n-2)(n-3)...(2)(1) \tag{4.3}$$

CHAPTER 4. SPECIAL FUNCTIONS

If we were to multiply both sides by n, we would just end up with $n!$. Then, multiplying that by $n+1$ gives us $(n+1)!$. Doing this repeatedly lets us derive the functional equation of $\Gamma(n)$.

$$n\Gamma(n) = \Gamma(n+1) \tag{4.4}$$

This can also be proven using the integral representation.

$$\Gamma(z+1) = \int_0^\infty t^z e^{-t}\, dt \tag{4.5}$$

Using integration by parts, we obtain:

$$\Gamma(z+1) = t^z e^{-t}\_0^\infty + \int_0^\infty z t^{z-1} e^{-t}\, dt \tag{4.6}$$

The first term evaluates to 0. Moving the z outside of the integral, we get,

$$\Gamma(z+1) = z \int_0^\infty t^{z-1} e^{-t}\, dt \tag{4.7}$$

This integral is exactly $\Gamma(z)$; therefore, we write the functional equation for gamma,

$$\Gamma(z+1) = z\Gamma(z) \tag{4.8}$$

Another important relationship comes from Leonhard Euler. It relates the gamma function to our regular trigonometric functions. It states,

> **Theorem 4.1.1: Euler Reflection Idenitity**
>
> $$\Gamma(x)\Gamma(1-x) = \frac{\pi}{\sin \pi x} \tag{4.9}$$

Example 4.1.1.

$$\int_0^\infty e^{-x^n}\, dx \tag{4.10}$$

First, let

$$u = x^n \tag{4.11}$$

$$du = nx^{n-1}\, dx \tag{4.12}$$

$$dx = \frac{u^{\frac{1-n}{n}}}{n}\, du \tag{4.13}$$

Therefore, substituting this back in, we get,

$$\frac{1}{n}\int_0^\infty e^{-u} u^{\frac{1}{n}-1}\, du \tag{4.14}$$

Using our integral representation of Γ, we find that

$$\frac{1}{n}\int_0^\infty e^{-u} u^{\frac{1}{n}-1}\, du = \frac{1}{n}\Gamma(\frac{1}{n}) \tag{4.15}$$

Using the functional equation, this can be simplified to

$$= \Gamma(\frac{1}{n}+1) \tag{4.16}$$

Lastly, combining it into one term, we get the final answer to the integral.

$$\int_0^\infty e^{-x^n}\, dx = \Gamma(\frac{n+1}{n}) \tag{4.17}$$

4.2 Beta Function

Now, in this section, we will talk about the Beta Function. In general, it is very similar in nature to the Gamma Function, except that this time it depends on two variables instead of one. There are a few different representations of the function based on whether or not you use trigonometric functions in the integral. The first representation looks rather similar to the gamma function integral representation.

Definition 4.2.1: First Beta Integral Definition

$$B(x,y) = \int_0^1 t^{x-1}(1-t)^{y-1} \, dy \qquad (4.18)$$

Then, you can make substitutions to turn the integral into a trigonometric one. The resulting Beta Function representation is

---
Definition 4.2.2: Second Beta Integral Definition

$$B(x,y) = 2\int_0^{\pi/2} \cos^{2x-1}\theta \sin^{2y-1}\theta \, d\theta \qquad (4.19)$$

If you substitute

$$t = \tan^2\theta \qquad (4.20)$$

into this integral, then you will arrive at another representation.

> **Definition 4.2.3: Third Beta Integral Definition**
>
> $$\int_0^\infty \frac{t^{x-1}}{(1+u)^{x+y}}\, dt \qquad (4.21)$$

Also, it can be expressed in the form of gamma functions.

> **Definition 4.2.4: Gamma-Beta Representation**
>
> $$B(x,y) = \frac{\Gamma(x)\Gamma(y)}{\Gamma(x+y)} \qquad (4.22)$$

Consider the following example.

# CHAPTER 4. SPECIAL FUNCTIONS

Example 4.2.1.

$$I = \int_0^\infty \frac{\ln x}{1+x^n}\,dx \qquad (4.23)$$

First make the substitution,

$$u = x^n \qquad (4.24)$$

$$du = nx^{n-1}\,dx \qquad (4.25)$$

$$dx = \frac{u^{\frac{1-n}{n}}}{n}\,du \qquad (4.26)$$

Substituting this back in, we get

$$\frac{1}{n^2}\int_0^\infty \frac{\ln u(u^{\frac{1-n}{n}})}{1+u}\,du \qquad (4.27)$$

This looks conspicuously like the third representation of the Beta Function. However, there is an additional $\ln u$ term. Where did this come from though? Well, if you remember back to the section on Feynman's Technique, one of the strategies was setting the parameter to be the exponent on x. That way, when you took the partial derivative, it spit out a $\ln x$ term. This is the exact same thing, except we are taking the partial derivative of the Beta Function.

Since u is just a dummy variable, we can change it back to x without it affecting the evaluation of the integral.

$$\frac{1}{n^2}\int_0^\infty \frac{\ln x(x^{\frac{1-n}{n}})}{1+x}\,dx \qquad (4.28)$$

Since we have determined that this is just the partial derivative of the Beta Function. Using the exponents, we know that the arguments are $B(x, 1-x)$. Using the gamma representation of the function, we can write that,

$$I = \frac{1}{n^2} \frac{\partial}{\partial x} B(x, 1-x) \tag{4.29}$$

$$I = \frac{1}{n^2} \frac{\partial}{\partial x} \frac{\Gamma(x)\Gamma(1-x)}{\Gamma(x+1-x)} \tag{4.30}$$

The bottom simplifies to $\Gamma(1)$, which equals 1. The top can be simplified using the Euler Reflection Formula.

$$I = \frac{1}{n^2} \frac{\partial}{\partial x} (\pi \csc \pi x)|_{x=\frac{1}{n}} \tag{4.31}$$

By taking the partial derivative and evaluating it at $x = \frac{1}{n}$, we arrive at the final answer to the integral.

$$I = \int_0^\infty \frac{\ln x}{1+x^n} dx = -\frac{\pi^2}{n^2} \cot \frac{\pi}{n} \csc \frac{\pi}{n} \tag{4.32}$$

4.3 Polygamma Function

The polygamma function is very similar to the gamma function because it is simply the derivative of the natural logarithm of the gamma function. We denote it as *psi*, with the superscript denoting how many derivatives have been taken.

> **Definition 4.3.1: Polygamma Function**
>
> $$\psi^{(n)}(z) = \frac{d^{n+1}}{dz^{n+1}}(\ln \Gamma(z)) \qquad (4.33)$$
>
> Another representation is found using infinite sums.
>
> $$\psi^{(n)}(z) = (-1)^{n+1} n! \sum_{k=0}^{\infty} \frac{1}{(k+z)^{n+1}} \qquad (4.34)$$

Example 4.3.1.

$$\int_0^1 \frac{\ln x}{1+x+x^2} \, dx \tag{4.35}$$

First, note that

$$(1+x+x^2)(1-x) = (1-x^3) \tag{4.36}$$

Now, we can multiply the top and bottom of the integral by $(1-x)$.

$$I = \int_0^1 \frac{(1-x)\ln x}{1-x^3} \, dx \tag{4.37}$$

Next, distribute the top and split it into two separate integrals.

$$I = \int_0^1 \frac{\ln x}{1-x^3} \, dx - \int_0^1 \frac{x \ln x}{1-x^3} \, dx \tag{4.38}$$

Substitute into the Maclaurin Series for both integrals.

$$\frac{1}{1-x^3} = \sum_{n=0}^{\infty} x^{3n} \tag{4.39}$$

and

$$\frac{x}{1-x^3} = \sum_{n=0}^{\infty} x^{3n+1} \tag{4.40}$$

Writing these into the integrals gives

$$I = \int_0^1 \ln x \sum_{n=0}^{\infty} x^{3n} \, dx - \int_0^1 \ln x \sum_{n=0}^{\infty} x^{3n+1} \, dx \tag{4.41}$$

In a previous section, we evaluated similar integrals by using integration by parts. Doing the same process turns the

CHAPTER 4. SPECIAL FUNCTIONS

integral into two sums.

$$I = \sum_{n=0}^{\infty} \frac{-1}{(3n+1)^2} - \frac{-1}{(3n+2)^2} \qquad (4.42)$$

Simplifying these sums,

$$I = \zeta(2) - \sum_{n=0}^{\infty} \frac{1}{(3n+1)^2} - \sum_{n=0}^{\infty} \frac{-1}{(3n)^2} - \sum_{n=0}^{\infty} \frac{1}{(3n+1)^2} \qquad (4.43)$$

$$I = \frac{8\zeta(2)}{9} - 2\sum_{n=0}^{\infty} \frac{1}{(3n+1)^2} \qquad (4.44)$$

You may be wondering what $\zeta(2)$ is. Well, we will talk about it in the next section, so it can't hurt to get a little bit of exposure now.

From the second term, factor out a $\frac{1}{9}$.

$$I = \frac{8\zeta(2)}{9} - \frac{2}{9}\sum_{n=0}^{\infty} \frac{1}{(\frac{n}{3}+1)^2} \qquad (4.45)$$

Now, it can be simply evaluated using the polygamma function. Since we can't find a closed-form for it, the best we can do is express it as

$$I = \frac{8\zeta(2)}{9} - \frac{2}{9}\psi^{(1)}\left(\frac{1}{3}\right) \qquad (4.46)$$

Lastly, using the fact that $\zeta(2) = \frac{\pi^2}{6}$, we can write the final answer to the integral.

$$\int_0^1 \frac{\ln x}{1+x+x^2}\, dx = \frac{4\pi^2}{27} - \frac{2}{9}\psi^{(1)}\left(\frac{1}{3}\right) \qquad (4.47)$$

Example 4.3.2.

$$\int_0^\infty \frac{\sin x}{\sinh x} \frac{\cos x}{\cosh x} dx \qquad (4.48)$$

By using the expansions of $\sinh x$ and $\cosh x$,

$$= \int_0^\infty \frac{\sin 2x}{\frac{e^{2x}-e^{-2x}}{2}} dx \qquad (4.49)$$

$$= 2\int_0^\infty \frac{\sin 2x}{e^{2x}-e^{-2x}} dx \qquad (4.50)$$

Let $u = 2x$, $du = 2x\,dx$ Since u is a dummy variable, transform it back to x

$$= \int_0^\infty \frac{sinx}{e^x - e^{-x}} dx \qquad (4.51)$$

Using $\sin x = \Im(e^{ix})$ gives

$$= \Im \int_0^\infty \frac{e^{ix}}{e^x - e^{-x}} dx \qquad (4.52)$$

$$= \Im \int_0^\infty \frac{e^{-x}e^{ix}}{e^{-x}(e^x - e^{-x})} dx \qquad (4.53)$$

$$= \Im \int_0^\infty \frac{e^{-x(1-i)}}{1 - e^{-2x}} \qquad (4.54)$$

By using,

$$\frac{1}{1-x} = \sum_{k=0}^\infty x^k \qquad (4.55)$$

$$\implies \frac{1}{1-e^{-2x}} = \sum_{k=0}^\infty e^{-2kx} \qquad (4.56)$$

CHAPTER 4. SPECIAL FUNCTIONS

Substituting this into the integral gives

$$I = \Im \int_0^\infty e^{-x(1-i)} \sum_{k=0}^\infty e^{-2kx} \, dx \qquad (4.57)$$

Combining the e terms,

$$= \Im \int_0^\infty \sum_{k=0}^\infty e^{-x(1+2k-i)} \, dx \qquad (4.58)$$

Interchanging the sum and integral:

$$\Im \sum_{k=0}^\infty \int_0^\infty e^{-x(1+2k-i)} \, dx \qquad (4.59)$$

$$\Im \sum_{k=0}^\infty (e^{-x(1+2k-i)})|_0^\infty \qquad (4.60)$$

$$= \Im \sum_{k=0}^\infty \frac{1}{1+2k-i} \qquad (4.61)$$

Multiply by the conjugate to the top and bottom.

$$= \Im \sum_{k=0}^\infty \frac{1+2k+i}{(1+2k)^2 - i^2} \qquad (4.62)$$

Factor the bottom and take the imaginary part of the sum.

$$= \sum_{k=0}^\infty \frac{1}{(2k+1+i)(2k+1-i)} \qquad (4.63)$$

$$= \frac{1}{2i} \sum_{k=0}^\infty \frac{1}{2k+1-i} - \frac{1}{2k+1+i} \qquad (4.64)$$

$$= \frac{1}{4i} \sum_{k=0}^{\infty} \frac{1}{k + \frac{1-i}{2}} - \frac{1}{k + \frac{1+i}{2}} \quad (4.65)$$

Recall the definition of the digamma function

$$\psi(z_1) - \psi(z_2) = \sum_{k=0}^{\infty} \frac{-1}{k + z_1} + \frac{1}{k + z_2} \quad (4.66)$$

$$I = \frac{1}{4i}(\psi(1 - \frac{1-i}{2}) - \psi(\frac{1-i}{2})) \quad (4.67)$$

Recall

$$\psi(1-z) - \psi(z) = \pi \cot(\pi z) \quad (4.68)$$

$$\implies \frac{1}{4i}(\psi(1 - \frac{1-i}{2}) - \psi(\frac{1-i}{2})) = \frac{1}{4i}\pi \cot(\frac{\pi}{2} - \frac{i\pi}{2}) \quad (4.69)$$

$$= \frac{\pi}{4i} \tan(\frac{i\pi}{2}) \quad (4.70)$$

$$= \frac{\pi}{4i}(i \tanh(\frac{\pi}{2})) \quad (4.71)$$

$$= \frac{\pi}{4} \tanh(\frac{\pi}{2}) \quad (4.72)$$

$$\int_0^{\infty} \frac{\sin x}{\sinh x} \frac{\cos x}{\cosh x} dx = \frac{\pi}{4} \tanh(\frac{\pi}{2}) \quad (4.73)$$

4.4 Zeta Function

First developed by Bernhard Riemann in his paper titled "On the Number of Primes Less Than a Given Magnitude", he originally used and proved its usefulness in determining the distribution of prime numbers. One specific case of the Riemann Zeta function is known as the Basel Problem. The problem asks for a closed-form evaluation of the famous infinite series,

$$\sum_{n=1}^{\infty} \frac{1}{n^2} \qquad (4.74)$$

It was first posed in 1650 by Pietro Mengoli and was eventually solved 84 years later by none other than Leonhard Euler. This question was then expanded to, "what about numbers other than two?". In this question lies the birth of the Riemann Zeta Function's first representation.

> **Definition 4.4.1: First Zeta Function Definition**
>
> $$\zeta(z) = \sum_{n=1}^{\infty} \frac{1}{n^z} \qquad (4.75)$$

$\zeta(2)$, the solution to the Basel Problem has been famously shown to equal $\frac{\pi^2}{6}$. Only the even values of ζ have been proven to have a closed form. $\zeta(3)$, $\zeta(5)$... have been shown to be irrational. Also, for the analytic continuation of the function, the integral representation has been derived.

> **Definition 4.4.2: Zeta Integral Representation**
>
> $$\zeta(z) = \frac{1}{\Gamma(z)} \int_0^\infty \frac{x^{z-1}}{e^x - 1} dx \qquad (4.76)$$

This equation has also appeared outside of pure mathematics. In quantum mechanics, certain integrals will take this form, which requires evaluation using the Zeta Function.

Additionally, there exists the Dirichlet Eta Function, which is very similar to the Zeta Function, with the exception that the signs are alternating instead of being strictly positive. The infinite series representation of it is given by,

> **Definition 4.4.3: Dirichlet Eta Function**
>
> $$\eta(z) = \sum_{n=1}^\infty \frac{(-1)^{n+1}}{n^z} \qquad (4.77)$$

Leonhard Euler, in 1749, showed that it is directly related to the Zeta Function via the following equation.

> **Theorem 4.4.1: Eta-Zeta Relationship**
>
> $$\eta(z) = (1 - 2^{1-z})\zeta(z) \qquad (4.78)$$

Now, we can see exactly when and how this mysterious function appears in the evaluation of integrals.

CHAPTER 4. SPECIAL FUNCTIONS

Example 4.4.1.

$$\int_0^1 (\tanh^{-1} x)^n \, dx \qquad (4.79)$$

The first step is to rewrite $\tanh^{-1} x$ as its natural logarithm counterpart.

$$I = \frac{1}{2^n} \int_0^1 \ln^n\left(\frac{1-x}{1+x}\right) dx \qquad (4.80)$$

Substitute x for $-x$. Next, let

$$u = \frac{1-x}{1+x} \qquad (4.81)$$

$$x = \frac{1-u}{1+u} \qquad (4.82)$$

$$dx = -\frac{2}{(1+u)^2} \, du \qquad (4.83)$$

Substitute this back into the integral and simplify.

$$I = \left(\frac{1}{2}\right)^n \int_0^1 \ln^n u \left(-\frac{2}{(1+u)^2}\right) du \qquad (4.84)$$

Cancel out the -2 with the fraction out front, and move it to the numerator by flipping the sign of the exponent. Since the sign is alternating, add a $(-1)^n$ term in front of the integral.

$$(-1)^n 2^{1-n} \int_0^1 \frac{\ln^n u}{(1+u)^2} \, du \qquad (4.85)$$

Now, make the substitution

$$\ln u = -t \qquad (4.86)$$

$$u = e^{-t} \tag{4.87}$$

$$du = -e^{-t} \, dt \tag{4.88}$$

Substituting this back in, we get a second $(-1)^n$ term. Since both terms will either be positive or both terms will be negative, we can get rid of them altogether.

$$\frac{1}{2^{n-1}} \int_0^\infty \frac{t^n e^{-t}}{(1+e^{-t})^2} \, dt \tag{4.89}$$

Next, we can expand the bottom term using an infinite series.

$$I = \frac{1}{2^{n-1}} \int_0^\infty t^n e^{-t} \sum_{k=1}^\infty (-1)^{k+1} k e^{-t(k-1)} \, dt \tag{4.90}$$

Switch the order

$$I = \frac{1}{2^{n-1}} \sum_{k=1}^\infty (-1)^{k+1} k \int_0^\infty t^n e^{-tk} \, dt \tag{4.91}$$

Make the substitution,

$$u = tk \tag{4.92}$$

$$du = k \, dt \tag{4.93}$$

Then, the integral becomes

$$\frac{1}{2^{n-1}} \sum_{k=1}^\infty (-1)^{k+1} k \frac{1}{k^{n+1}} \int_0^\infty u^n e^{-u} \, du \tag{4.94}$$

Apply the definition of the gamma function.

$$I = \frac{\Gamma(n+1)}{2^{n-1}} \sum_{k=1}^\infty \frac{(-1)^{k+1}}{k^n} \tag{4.95}$$

CHAPTER 4. SPECIAL FUNCTIONS

Since $(-1)^{k+1} = (-1)^{k-1}$, the sum is precisely equal to the Dirichlet Eta Function. Therefore, substituing it in, we get the final answer to the integral

$$\int_0^1 (\tanh^{-1} x)^n \, dx = \frac{\Gamma(n+1)\eta(n)}{2^{n-1}} \qquad (4.96)$$

Chapter 5

Contour Integration

Now, we have arrived at our last and arguably most difficult section. Before we can jump in, it would be wise to get some basics of complex analysis. They are going to be crucial skills that allow us to solve these integrals. We have to be able to perform a lot of manipulation that you only really see with complex functions; thus, many of them are very new. Without any further delay, let's talk about the precursors of complex analysis.

5.1 Basics of Complex Analysis

This precursor section will explain the main points of complex analysis that are required for evaluating integrals using contour integration. They will be split up by topic and will be short and straight to the point.

> **Definition 5.1.1: Points in the Complex Plane**
>
> or a complex number z, it can be written in the form
>
> $$z = a + bi \tag{5.1}$$
>
> where $i = \sqrt{-1}$

It can be plotted on the complex plane where the y-axis becomes the imaginary coordinate, and the x-axis becomes the real coordinate.

To find the "length" of the coordinate in the complex plane, you can take the absolute value. For,

$$z = x + iy \tag{5.2}$$

$$|z| = \sqrt{x^2 + y^2} \tag{5.3}$$

Then, using the absolute values, we can form the triangle inequalities.

$$-|z| \leq \Re(z) \leq |\Re(z)| \leq |z| \tag{5.4}$$

$$-|z| \leq \Im(z) \leq |\Im(z)| \leq |z| \tag{5.5}$$

CHAPTER 5. CONTOUR INTEGRATION

$\Re(z)$ is the "real part of z". For this case, it equals the x-value of z. Conversely, $\Im(z)$ is the imaginary part of z, the y-value.

> **Definition 5.1.2: Complex Conjugate**
>
> The complex conjugate essentially reflects the point in the complex plane over the x-axis. If
>
> $$z = x + iy \qquad (5.6)$$
>
> then
>
> $$z^* = x - iy \qquad (5.7)$$

Multiplying a complex number by its conjugate gives

$$z^*z = |z|^2 \qquad (5.8)$$

The complex conjugate operator is also linear. In other words,

$$(z+w)^* = z^* + w^* \qquad (5.9)$$

The same property applies to multiplication as well.

$$(zw)^* = z^*w^* \qquad (5.10)$$

The first basic operation is addition. Much like regular addition, it is done term by term.

$$z_1 + z_2 = (x_1 + x_2) + (y_1 + y_2)i \qquad (5.11)$$

Second, is subtraction. The process is almost completely the same as addition.

$$z_1 - z_2 = (x_1 - x_2) + (y_1 - y_2)i \quad (5.12)$$

Third, is multiplication. For this one, you have to use FOIL, then group together the real and imaginary terms.

$$z_1 z_2 = (x_1 x_2 - y_1 y_2) + i(y_1 x_2 + x_1 y_2) \quad (5.13)$$

Lastly, is division.
$$\frac{z_1}{z_2} = \frac{z_1 z_2^*}{|z_1|^2} \quad (5.14)$$

From the operations, we can derive numerous triangle inequalities that help us relate and compare different complex quantities in terms of their absolute values. Consider a list of some of them.

$$|z + w|^2 = |z| + 2\Re(zw^*) + |w|^2 \quad (5.15)$$
$$|z + w| \leq |z| + |w| \quad (5.16)$$
$$|z - w| \leq |z| + |w| \quad (5.17)$$
$$||z| - |w|| \leq |z - w| \quad (5.18)$$
$$||z| - |w|| \leq |z + w| \quad (5.19)$$

Now, we can start to look at specific functions and how they behave in the complex plane. For example, we can look at the complex analog of the exponential function.

> **Definition 5.1.3: Complex Exponential Function**
>
> $$|e^z| \leq \sum_{n=0}^{\infty} \frac{|z|^n}{n!} = e^{|z|} \qquad (5.20)$$

A special case regarding the exponential function is that of Euler's Formula. It states that

We can look at another way to write a complex number other than splitting its real and complex parts and adding them separately.

> **Definition 5.1.4: Complex Polar Representation**
>
> Any point, $z = x + iy$, can be written in terms of polar coordinates by using
>
> $$z = |z|e^{i\theta} \tag{5.21}$$

Instead of writing out $|z|$ every time, we simplify it by calling $|z| = R$. Therefore, the polar representation of a complex point becomes,

$$z = Re^{i\theta} \tag{5.22}$$

The argument of z, $arg(z)$, is the angle, in radians, between the positive x-axis and the line $z = Re^{i\theta}$. For θ between 5π), $arg(z) = \theta$. For any angle outside of this interval, we denote it arg(z). $\quad arg(z) = \theta + 2\pi k \tag{5.23}$

To take just the argument between $[-\pi, \pi$

Intuitively speaking, since the exponential function can take on the exact same value at different angles, the natural logarithm could yield different values at one value of z. Doing this would violate the definition of a function, so to get around this, we must define and use the principal branch of the logarithm function.

$$\ln(z) = \ln(|z|) + i\arg(z) \tag{5.24}$$

5.2 Contour Integrals

We have finally made it to our last section, contour integration. First, we need to look at some terminology that helps to explain the upcoming theorems and definitions.

A function, $f(z)$, is holomorphic if and only if it is complex differentiable, or its derivative exists over the entirety of the complex plane. If this is not true, then the function has a "pole". Each pole has its own order based on the power of z in the denominator of the fraction or the general behavior of the function at a point. Essential singularities are those that are of infinite order, such as $(e^{\frac{1}{z}}, 0)$.

An analytic function is a function whose values are locally equal to those of its Taylor Series (infinite series centered around a point).

Important Rule:

Any holomorphic function is also analytic at every point.

Moving along, we can look at some contour integral evaluation techniques. First up is Cauchy's Integral Theorem. This states that

CHAPTER 5. CONTOUR INTEGRATION

> **Theorem 5.2.1: Cauchy's Integral Theorem**
>
> If $f(z)$ is holomorphic over a contour C, then
>
> $$\oint_C f(z)\, dz = 0 \qquad (5.25)$$

Let
$$f(z) = u + iv \qquad (5.26)$$
$$dz = dx + i\,dy \qquad (5.27)$$

Substituting this into the contour integral,

$$\oint_C f(z)\, dz = \oint_C (u + iv)(dx + i\, dy) \qquad (5.28)$$

$$= \oint_C (u\, dx - v\, dy) + i \oint_C (v\, dx + u\, dy) \qquad (5.29)$$

Now, we can use Green's Theorem to rewrite the contour integral as a double integral over a general region, and apply the Cauchy-Riemann equations that the function must satisfy in order to be holomorphic. We find that both integrals evaluate to 0, proving that our entire integral collapses to 0 if $f(z)$ is holomorphic.

Another useful trick is to split one contour into many. Suppose you have a contour C. You can split it up into the integrals over C_1, C_2, and so on. This creates the definition,

$$\oint_C = \oint_{C_1} + \oint_{C_2} + \ldots \qquad (5.30)$$

Another useful theorem is Cauchy's Residue Theorem. It states that

> **Theorem 5.2.2: Cauchy's Residue Theorem**
>
> If there is exactly one pole, $z = z_0$ enclosed in a contour , C, then
>
> $$\oint_C f(z)\,dz = 2\pi i \text{Res}(f(z), z_0) \qquad (5.31)$$

If there is more than one pole, then you would add up the residues at each of those points. But how do we find the residues? Well, there is a formula that is based on the order of the singularity.

$$\text{Res}(f(z), z_0) = \frac{1}{(n-1)!} \lim_{z \to z_0} \frac{d^{n-1}}{dz^{n-1}}((z-z_0)^n f(z_0)) \quad (5.32)$$

If the pole lands on the x or y-axis, often times a contour will avoid it. Calculating residues at those points is difficult/impossible in most cases. Consider the contour

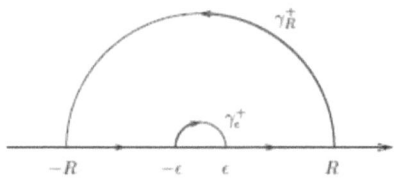

Figure 5.1: Indented Contour

Since there was a pole at $z = 0$, the contour avoided it. Therefore, combining the last couple pieces of information, we can write what the contour integral of this function would look like. Of course, I didn't actually provide the function, so this is just a guess of sorts. All that we can assume is that the integral of $f(x)$ is taken over $(-\infty, \infty)$. Since the function is holomorphic over the contour,

$$\oint_C f(z)\, dz = \int_\epsilon^R + \int_{\gamma_R^+} + \int_{-R}^{-\epsilon} + \int_{\gamma_\epsilon^+} = 0 \qquad (5.33)$$

How do we evaluate all of the integrals though? First, take the limit as $\epsilon \to 0$ and $R \to \infty$, then, we can write

$$\oint_C f(z)\, dz = \int_{-\infty}^{\infty} f(x)\, dx + \int_{\gamma_R^+} + \int_{\gamma_\epsilon^+} = 0 \qquad (5.34)$$

We were able to transform the integral over $(-\infty, \infty)$ back to x since it is entirely on the real part of the complex plane. To evaluate the two contours, we can employ a bevy of techniques.

The first technique we can use is the Estimation Lemma (ML Inequality).

Lemma 5.2.1: Estimation Lemma

$$\left| \int_\gamma f(z)\, dz \right| \leq ML \qquad (5.35)$$

where M is the maximum of $|f(z)|$, and L is the length of the contour γ. Then, you can use the following identity of

integrals.
$$\left|\int_\gamma f(z)\,dz\right| \leq \int_\gamma |f(z)|\,dz \qquad (5.36)$$

The next step is to parameterize the integral. For example, let
$$z = Re^{i\theta} \qquad (5.37)$$
$$dz = iRe^{i\theta}\,d\theta \qquad (5.38)$$

You can also make use of the triangle inequalities to obtain the integral that is less than some function of R, where you can take the limit as $R \to \infty$, ideally showing that it either converges to a value or collapses to zero.

Then, you would do the exact same thing, except you would parameterize the other curve using ϵ, and once you get an expression, you can take $\epsilon \to 0$. Then, once you obtain those results, you can incorporate them into the equation from above and solve for the integral that you are looking for.

Additionally, you can use Jordan's Lemma; however, this only works with a semi-circular contour. Then, you can apply the following formula

Lemma 5.2.2: Jordan's Lemma

$$\left|\int_\gamma e^{iaz} f(z)\,dz\right| \leq \frac{\pi}{a} \max |f(z)| \qquad (5.39)$$

Example 5.2.1.

$$\int_{-\infty}^{\infty} \frac{1}{x^2+1}\, dx \qquad (5.40)$$

To integrate this, we will utilize a semi-circular contour like such.

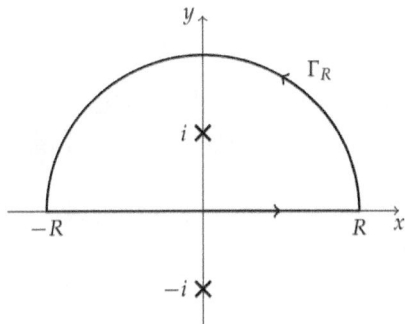

Figure 5.2: Semi-circular Contour

Since the contour contains one single pole at $z = i$, we can use the Residue Theorem.

$$\oint_C f(z)\, dz = 2\pi i \operatorname{Res}(f(z), i) = \int_\Gamma + \int_{-\infty}^{\infty} f(x)\, dx \qquad (5.41)$$

To evaluate the residue, first factor $f(z)$

$$f(z) = \frac{1}{(z+i)(z-i)} \qquad (5.42)$$

These are both single-ordered poles, so finding the residue looks like

$$\operatorname{Res}(\frac{1}{(z+i)(z-i)}, i) = \lim_{z \to i}(z-i)(\frac{1}{(z-i)(z+i)}) \qquad (5.43)$$

Substitute in i and cancel out the like terms. Therefore,

$$Res(\frac{1}{(z+i)(z-i)}, i) = \frac{1}{2i} \qquad (5.44)$$

Multiplying this by $2\pi i$ gives

$$\oint_C f(z)\, dz = \pi = \int_\Gamma + \int_{-\infty}^{\infty} f(x)\, dx \qquad (5.45)$$

Now, we will determine what the value of the integral over Γ converges to using the Estimation Lemma. First make the substitution,

$$z = Re^{i\theta} \qquad (5.46)$$

$$dz = iRe^{i\theta}\, d\theta \qquad (5.47)$$

Now, substitute this back into the lemma.

$$|\int_\Gamma \frac{1}{R^2 e^{2i\theta}}| \leq ML \qquad (5.48)$$

Since this is just a semi-circle, $L(\Gamma) = \pi R$
On the side, take

$$|\frac{1}{z^2+1}| = \frac{1}{|z^2+1|} \qquad (5.49)$$

$$= \frac{1}{|z^2 - (-1)|} \qquad (5.50)$$

$$\leq \frac{1}{||z|^2 - 1|} \qquad (5.51)$$

$$= \frac{1}{R^2 - 1} \qquad (5.52)$$

CHAPTER 5. CONTOUR INTEGRATION

Therefore, multiplying this by the length of Γ, we find that

$$\left| \int_\Gamma \frac{1}{z^2+1} \right| \leq \frac{\pi R}{R^2-1} \tag{5.53}$$

Now, take the limit as $R \to \infty$. We find that,

$$\left| \int_\Gamma \frac{1}{z^2+1} \right| \leq 0 \tag{5.54}$$

Therefore,

$$\int_\Gamma = 0 \tag{5.55}$$

Since we have all of our parts evaluated, we can find out what the integral equals. Going back to the original equation, we can substitute in 0 for the integral over Γ.

$$\oint_C f(z)\, dz = \pi = \int_{-\infty}^{\infty} f(x)\, dx \tag{5.56}$$

$$\pi = \int_{-\infty}^{\infty} f(x)\, dx \tag{5.57}$$

In the end, we wind up with our final answer for the integral.

$$\int_{-\infty}^{\infty} \frac{1}{x^2+1}\, dx = \pi \tag{5.58}$$

Example 5.2.2.

$$I = \int_0^\infty \frac{2\cos x \ln x + \pi \sin x}{x^2 + 4} \, dx \tag{5.59}$$

First, let

$$f(z) = \frac{e^{iz} \ln z}{z^2 + 4} \tag{5.60}$$

Utilize the contour, Letting $\epsilon \to 0$, and letting $R \to \infty$,

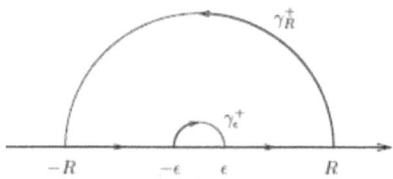

Figure 5.3: Indented Semi-circular Contour

$$\oint_C f(z) \, dz = \int_\Gamma + \int_\gamma + I_1 + \int_{-\infty}^0 \frac{e^{ix} \ln x}{x^2 + 4} \, dx \tag{5.61}$$

Using the fact that

$$\ln(-x) = \ln(-1) + \ln x = i\pi + \ln x \tag{5.62}$$

In the integral,

$$\int_{-\infty}^0 \frac{e^{ix} \ln x}{x^2 + 4} \, dx \tag{5.63}$$

Let $x \to -x$,

$$\int_{-\infty}^0 \frac{e^{-ix} \ln(-x)}{x^2 + 4} \, dx \tag{5.64}$$

CHAPTER 5. CONTOUR INTEGRATION

We can write it as

$$\int_0^\infty \frac{e^{-ix} \ln x + i\pi e^{-ix}}{x^2 + 4} \, dx \tag{5.65}$$

Let this integral be I_2 and add I_1 and I_2.

$$I_1 + I_2 = \int_0^\infty \frac{e^{ix} + e^{-ix} \ln x + i\pi e^{-ix}}{x^2 + 4} \, dx \tag{5.66}$$

Using the identity derived from Euler's Formula

$$2 \cos x = e^{ix} + e^{-ix} \tag{5.67}$$

and the formula itself,

$$i\pi e^{-ix} = i\pi \cos x + \pi \sin x \tag{5.68}$$

we can substitute this into the integral.

$$I_1 + I_2 = \int_0^\infty \frac{2 \cos x \ln x + \pi \sin x}{x^2 + 4} \, dx + i\pi \int_0^\infty \frac{\cos x}{x^2 + 4} \, dx \tag{5.69}$$

Therefore, we arrive at

$$\oint_C f(z) \, dz = \int_\Gamma + \int_\gamma + I + i\pi \int_0^\infty \frac{\cos x}{x^2 + 4} \, dx \tag{5.70}$$

Now we can use the residue theorem since we have a pole at $z = 2i$.

$$\oint_C f(z) \, dz = 2\pi i \operatorname{Res}(f(z), 2i) \tag{5.71}$$

$$\oint_C f(z) \, dz = \frac{\pi e^{-2} \ln 2i}{2} \tag{5.72}$$

$$\ln 2i = \frac{i\pi}{2} + \ln 2 \tag{5.73}$$

We can substitute this into the integral yet again.

$$\oint_C f(z)\, dz = \frac{\pi \ln 2}{2e^2} + \frac{\pi^2}{4e^2} i = I + \int_\Gamma + \int_\gamma \tag{5.74}$$

I will leave it to the reader to show that both of the latter two integrals converge. You will use the exact same method as in the previous example. In the end, you will end up with

$$\frac{\pi \ln 2}{2e^2} + \frac{\pi^2}{4e^2} i = I + i\pi \int_0^\infty \frac{\cos x}{x^2 + 4}\, dx \tag{5.75}$$

Factor out a π from the second term on the right-hand side.

$$\frac{\pi \ln 2}{2e^2} + i\pi \frac{\pi}{4e^2} = I + i\pi \int_0^\infty \frac{\cos x}{x^2 + 4}\, dx \tag{5.76}$$

From this, we can conclude the results of two integrals.

$$\int_0^\infty \frac{\cos x}{x^2 + 4}\, dx = \frac{\pi}{4e^2} \tag{5.77}$$

and our original integral we were trying to evaluate

$$\int_0^\infty \frac{2\cos x \ln x + \pi \sin x}{x^2 + 4}\, dx = \frac{\pi \ln 2}{2e^2} \tag{5.78}$$

Chapter 6

Section on Series

Considering that infinite series have come up repeatedly, it would be useful to quickly discuss them and present some methods for evaluating them.

6.1 Recognizing Functions

First, we can talk about looking at the Maclaurin series. This is probably the most intuitive of them all, for example, if you have

$$\sum_{k=0}^{\infty} \frac{(-1)^k}{2k+1} \qquad (6.1)$$

This looks a lot like the series expansion of $\arctan x$. Well, in fact, if you were to plug in 1 into the series expansion, you would end up with just that. Since we know that $\arctan(1) = \frac{\pi}{4}$, we know that this bizarre looking sum evaluates to

$$\sum_{k=0}^{\infty} \frac{(-1)^k}{2k+1} = \frac{\pi}{4} \qquad (6.2)$$

6.2 Derivatives and Integrals

Another possible method is to take a known infinite series and derive the function inside or even integrate it until you obtain the sum that you have. Then, take the derivative or integral of your function and plug in the value that you need to evaluate the sum.

Example 6.2.1.

$$\sum_{n=1}^{\infty} \frac{1}{n^2 2^n} \quad (6.3)$$

Using an integral, the sum equals

$$S = \sum_{n=1}^{\infty} \int_0^{1/2} \frac{x^{n-1}}{n} \, dx \quad (6.4)$$

Now, we can change the order of the sum/integral.

$$\int_0^{1/2} \sum_{n=1}^{\infty} \frac{x^{n-1}}{n} \, dx \quad (6.5)$$

Using our knowledge of infinite series expansions of functions, we can write

$$S = -\int_0^{1/2} \frac{\ln(1-x)}{x} \, dx \quad (6.6)$$

Next, use integration by parts.

$$S = -\ln x \ln(1-x)\big|_0^{1/2} + \int_0^{1/2} \frac{\ln x}{x-1} \, dx \quad (6.7)$$

CHAPTER 6. SECTION ON SERIES

Starting from the series expansion of $\frac{1}{1+x}$, we can integrate it, and substitute in $x \to x - 1$ to find the series expansion of $\ln x$.

$$\ln x = \sum_{n=0}^{\infty} \frac{(-1)^n (x-1)^{n+1}}{n+1} \tag{6.8}$$

Then, substitute this back into the integral.

$$S = -\ln^2\left(\frac{1}{2}\right) + \int_0^{1/2} \sum_{n=0}^{\infty} \frac{(-1)^n (x-1)^n}{n+1} \, dx \tag{6.9}$$

We can then switch the order of integral/sum once again and integrate the function.

$$S = -\ln^2\left(\frac{1}{2}\right) + \sum_{n=0}^{\infty} \frac{(-1)^n}{n+1} \left(\frac{(x-1)^{n+1}}{n+1}\bigg|_0^{1/2}\right) \tag{6.10}$$

Plugging in the bounds and shifting the index of the sum by 1 gives us,

$$S = -\ln^2\left(\frac{1}{2}\right) - \sum_{n=1}^{\infty} \frac{1}{n^2 2^n} + \sum_{n=0}^{\infty} \frac{1}{(n+1)^2} \tag{6.11}$$

The first sum is exactly the one we are looking for, so we can add it to the other side. If we move the index by 1 of the second sum to where it is at $n = 1$, we end up exactly with $\zeta(2)$. Therefore, our sum looks like

$$2S = -\ln^2 2 + \frac{\pi^2}{6} \tag{6.12}$$

Dividing everything by 2 gives us our final answer to the sum.

$$\sum_{n=1}^{\infty} \frac{1}{n^2 2^n} = \frac{\pi^2}{12} - \frac{1}{2}\ln^2 2 \tag{6.13}$$

6.3 Fourier Series

This section covers a very large subject with many different applications to physics, engineering, pure mathematics, etc. The Fourier Series is a sum of sine and cosine waves that, when added together with the right coefficients, spits out a function $f(x)$. Generally, the function is only defined over intervals of π. The function $f(x)$ is defined as

$$f(x) = a_0 + \sum_{n=1}^{\infty} A_n \cos(2\pi \frac{nx}{L}) + B_n \sin((2\pi \frac{nx}{L}) \quad (6.14)$$

L is the period during which you are working. For example, L could range from $(-\pi, \pi)$. The coefficients, a_n, b_n, and a_0, are what give the sum its characteristic look that allows it to resemble $f(x)$. They are found by taking three integrals.

$$a_0 = \frac{1}{L} \int_{-L}^{L} f(x)\, dx \quad (6.15)$$

$$a_n = \frac{1}{L} \int_{-L}^{L} f(x) \cos(2\pi \frac{nx}{L})\, dx \quad (6.16)$$

$$b_n = \frac{1}{L} \int_{-L}^{L} f(x) \sin(2\pi \frac{nx}{L})\, dx \quad (6.17)$$

Something to note is that depending on whether $f(x)$ is even or odd, that changes or simplifies the coefficients. For example, if $f(x)$ is an odd function, then integrating over a symmetric interval of an odd function will always be 0. Therefore, whenever $f(x)$ is odd,

$$a_n = 0 \quad (6.18)$$

This is useful for evaluating many different sums; however, it is limited to those containing trigonometric functions or those that can be easily expressed by terms of trigonometric functions.

6.4 Complex Analysis

Lastly, we look at different techniques that go hand-in-hand with what we looked at in the previous chapter. There are three main theorems that we can apply for a function $f(z)$. First, we will look at what happens if it is just $f(k)$ in the sum.

$$\sum_{-\infty}^{\infty} f(k) = -\sum Res(\pi \cot(\pi z) f(z), z_n) \qquad (6.19)$$

You have to take the sum of the residues at each value of z, z_0, z_1 until you get to z_n, your last pole.

Suppose you have $f(k)$ multiplied by an alternating term. Then, you can use this second theorem.

$$\sum_{-\infty}^{\infty} (-1)^k f(k) = -\sum \operatorname{Res}(\pi \csc(\pi z) f(z), z_n) \qquad (6.20)$$

The same principle applies. Take the sum of the residues at each of your poles. The only difference is that now we use $\csc(\pi z)$ instead of $\cot(\pi z)$.

The third theorem is that you can take the derivative of the function inside an complex-valued power series and plug in values to find related series so long as you are taking the derivative at a point that is within its convergence.

Example 6.4.1.

$$\sum_{-\infty}^{\infty} \frac{1}{k^2 + a^2} \qquad (6.21)$$

We can find the residues of $f(k)\pi \cot(\pi z)$. We find, then, that the sum of the residues.

$$-\sum Res = 2\frac{\pi \coth \pi a}{a} \qquad (6.22)$$

Since

$$\pi \cot(\pi z) \qquad (6.23)$$

has a pole at every integer k, we evaluate the residue and sum it from $(-\infty, \infty)$.

$$\sum Res(k) = \sum_{-\infty}^{\infty} \frac{1}{k^2 + a^2} \qquad (6.24)$$

Now that we have our components, we can add solve for the value of the sum.

$$\sum_{-\infty}^{\infty} \frac{1}{k^2 + a^2} = 2\frac{\pi \coth \pi a}{a} - \sum_{-\infty}^{\infty} \frac{1}{k^2 + a^2} \qquad (6.25)$$

Add the sum over and divide by 2 to get our final answer.

$$\sum_{-\infty}^{\infty} \frac{1}{k^2 + a^2} = \frac{\pi \coth \pi a}{a} \qquad (6.26)$$

Chapter 7

Applications

With learning all of these different techniques for solving abstract integrals, it is helpful to see exactly where these techniques are applied. A lot of them have very little purpose in the real world; however, there are many that do arise. From Physics, Statistics, Economics, and even more fields, integrals are showing up everywhere. A majority of those appearing in nature are based on a famous integral, but there is something about them that prevents a closed-form from existing. Therefore, we have to come up with numerical methods to solve them. Many famous integrals that were considered "impossible" arose from real-world problems that led to curiosity, which, in turn, led to discovery. If we can maintain this curiosity for mathematics, our society and future generations will be highly successful.

7.1 Physics

Physics is such a broad subject that it is not surprising that integrals appear at least somewhere. Not only do they appear somewhere, but they are appearing everywhere. Integrals appear in kinematics when finding displacement and the change in the velocity of a particle.

$$\int v\ dt = \Delta x \tag{7.1}$$

$$\int a\ dt = \Delta v \tag{7.2}$$

Also, it shows up when discussing the work-energy theorem and the impulse-momentum theorem. This could be useful in engineering, for example, in determining how much give a car should have so that the person inside does not experience a fatal force in the event of a car crash. If the car is too rigid, then the person will go flying through the windshield. If the car has too much give, then the person will be crushed. The job of the engineer is to find the happy medium of where it is safe to drive. They can achieve this by using the integral provided by the impulse-momentum theorem.

$$I = \int F\ dt \tag{7.3}$$

The four fundamental equations of electromagnetism, the Maxwell Equations, are also integrals. They arise as cases of Green's and Stokes theorems. They are useful for determining how the movement of a charged object creates a magnetic field and how there is no such thing as a magnetic

CHAPTER 7. APPLICATIONS   141

monopole. These are still used today in electrical engineering, especially in the development of smartphones, computers, televisions, and much more.

Pendulums are another great example. To find the period, we have to look at what is known as the eliptic integral.

$$F(\frac{\pi}{2}, k) = \int_0^{\pi/2} \frac{d\theta}{\sqrt{(1 - k^2 \sin^2 \theta)}} \qquad (7.4)$$

Recently, I wrote a paper examining the effects of air resistance on a simple pendulum. In there, I solved that integral to find the actual period of a pendulum, not just the approximation that every physics teacher gives you.

7.2 Differential Equations

Differential equations are the backbone of physics, mathematics, and even finance. Solving them often requires integrals. For example, the heat equation requires the use of Fourier transforms in order to solve it. The same goes for the wave equation, the Laplace Equation, and many more. What do all of these have in common? They all require the use of integrals. Integrals are literally everywhere.

Looking back after writing *Differential Equations*, I realized just how often integrals are used. I mention it briefly here, but I feel I really understated their actual value.

7.3 Statistics

One of the major topics in statistics is distributions. named after Gauss, the bell curve is given by the equation

$$f(x) = e^{-x^2} \qquad (7.5)$$

To find the probability of landing between two points on this curve, you would have to integrate it from point a to point b. This requires changing the integral from a single one to a double one, then switching the coordinates from cartesian to polar. I won't entirely solve it here, but it is a fun topic to research and learn about. The final answer is

$$\int_{-\infty}^{\infty} e^{-x^2} \, dx = \sqrt{\pi} \qquad (7.6)$$

In the end, we have shown how valuable integrals are to everything that we do in the world today. In almost every field, they can be applied and used to solve all kinds of problems. Their usefulness cannot be overstated.

Chapter 8

Additional Problems

Below are some problems that I feel are very valuable to your success solving integrals. All of these will utilize methods talked about in the previous chapters. Some integrals may require outside sources (looking up power series, Fourier series, specific constants, etc.).

Exercise 8.0.1.

$$\int x \cos x \, dx \tag{8.1}$$

Exercise 8.0.2.

$$\int x^2 e^x \, dx \tag{8.2}$$

Exercise 8.0.3.

$$\int \frac{1}{x^2 + 1} \, dx \tag{8.3}$$

Exercise 8.0.4.

$$\int \sin^3 x \, dx \tag{8.4}$$

Exercise 8.0.5.

$$\int \cos^5 x \, dx \tag{8.5}$$

Exercise 8.0.6.

$$\int \frac{1}{\sqrt{a^2 - x^2}} \, dx \tag{8.6}$$

Exercise 8.0.7.

$$\int \frac{1}{x \ln x} \, dx \tag{8.7}$$

Exercise 8.0.8.

$$\int \frac{x}{x^2 + 1} \, dx \tag{8.8}$$

Exercise 8.0.9.

$$\int \frac{1}{(x-1)(x+2)} \, dx \tag{8.9}$$

Exercise 8.0.10.

$$\int \frac{1}{x^3 - 1} \, dx \tag{8.10}$$

Exercise 8.0.11.

CHAPTER 8. ADDITIONAL PROBLEMS

Exercise 8.0.12.
$$\int \frac{x^2}{x^2+1}\,dx \qquad (8.11)$$

Exercise 8.0.13.
$$\int \frac{1}{(x^2+1)^2}\,dx \qquad (8.12)$$

Exercise 8.0.14.
$$\int \tan x\,dx \qquad (8.13)$$

Exercise 8.0.15.
$$\int \sec x\,dx \qquad (8.14)$$

Exercise 8.0.16.
$$\int \sin x \cos x\,dx \qquad (8.15)$$

Exercise 8.0.17.
$$\int \frac{1}{\sqrt{x^2+a^2}}\,dx \qquad (8.16)$$

Exercise 8.0.18.
$$\int x\sqrt{x+1}\,dx \qquad (8.17)$$

$$\int \frac{1}{x^2+\sqrt{x}}\,dx \qquad (8.18)$$

Exercise 8.0.19.
$$\int \frac{x+1}{x^2+3x+2}\,dx \tag{8.19}$$

Exercise 8.0.20.
$$\int \frac{1}{(x+1)\sqrt{x}}\,dx \tag{8.20}$$

Exercise 8.0.21.
$$\int \ln x\,dx \tag{8.21}$$

Exercise 8.0.22.
$$\int x\ln x\,dx \tag{8.22}$$

Exercise 8.0.23.
$$\int_0^1 x^n\,dx \tag{8.23}$$

Exercise 8.0.24.
$$\int_0^\infty e^{-ax}\,dx \tag{8.24}$$

Exercise 8.0.25.
$$\int_0^\infty x^n e^{-x}\,dx \tag{8.25}$$

Exercise 8.0.26.

CHAPTER 8. ADDITIONAL PROBLEMS 149

$$\int_0^{\pi/2} \sin^m x \cos^n x \, dx \qquad (8.26)$$

Exercise 8.0.27.

$$\int_0^1 x^{p-1}(1-x)^{q-1} \, dx \qquad (8.27)$$

Exercise 8.0.28.

$$\int_0^\infty \frac{x^{a-1}}{1+x} \, dx \qquad (8.28)$$

Exercise 8.0.29.

$$\int_0^\infty \frac{1}{(1+x^2)^2} \, dx \qquad (8.29)$$

Exercise 8.0.30.

$$\int_0^\infty \frac{\sin x}{x} \, dx \qquad (8.30)$$

Exercise 8.0.31.

$$\int \frac{\ln x}{1+x^2} \, dx \qquad (8.31)$$

Exercise 8.0.32.

$$\int_0^1 \frac{\ln(1+x)}{x} \, dx \qquad (8.32)$$

Exercise 8.0.33.

$$\int_0^1 \frac{1-x^n}{1-x}\,dx \qquad (8.33)$$

Exercise 8.0.34.

$$\int e^x \sin x\,dx \qquad (8.34)$$

Exercise 8.0.35.

$$\int e^{ax} \cos(bx)\,dx \qquad (8.35)$$

Exercise 8.0.36.

$$\int \frac{1}{x\sqrt{x^2-1}}\,dx \qquad (8.36)$$

Exercise 8.0.37.

$$\int \sqrt{a^2+x^2}\,dx \qquad (8.37)$$

Exercise 8.0.38.

$$\int \frac{x^2}{\sqrt{x^2+a^2}}\,dx \qquad (8.38)$$

Exercise 8.0.39.

$$\int \frac{\arctan x}{x}\,dx \qquad (8.39)$$

Exercise 8.0.40.

$$\int \arcsin x\,dx \qquad (8.40)$$

CHAPTER 8. ADDITIONAL PROBLEMS 151

Exercise 8.0.41.
$$\int \frac{1}{x^4+1}\,dx \qquad (8.41)$$

Exercise 8.0.42.
$$\int_0^{2\pi} \frac{1}{a+b\cos\theta}\,d\theta \qquad (8.42)$$

Exercise 8.0.43.
$$\int_0^{2\pi} \cos(n\theta)\,d\theta \qquad (8.43)$$

Exercise 8.0.44.
$$\int_0^\infty \frac{x}{e^x-1}\,dx \qquad (8.44)$$

Exercise 8.0.45.
$$\int_0^\infty \frac{x^{s-1}}{e^x-1}\,dx \qquad (8.45)$$

Exercise 8.0.46.
$$\int_0^\infty \frac{\sin(ax)}{x} e^{-bx}\,dx \qquad (8.46)$$

Exercise 8.0.47.
$$\int_0^\infty \frac{\cos(ax)-e^{-bx}}{x}\,dx \qquad (8.47)$$

Exercise 8.0.48.

$$\int_0^\infty \frac{\sin x}{x^p}\,dx \tag{8.48}$$

Exercise 8.0.49.

$$\int_0^\infty e^{-x^2}\,dx \tag{8.49}$$

Exercise 8.0.50.

$$\int_{-\infty}^\infty e^{-x^2}\,dx \tag{8.50}$$

Exercise 8.0.51.

$$\int_{-\infty}^\infty e^{-ax^2+bx}\,dx \tag{8.51}$$

Exercise 8.0.52.

$$\int_0^\infty x^{\nu-1} e^{-\beta x^\mu}\,dx \tag{8.52}$$

Exercise 8.0.53.

$$\int_0^\infty \frac{\cos x}{1+x^2}\,dx \tag{8.53}$$

Exercise 8.0.54.

$$\int_0^\infty \frac{\sin x}{x(1+x^2)}\,dx \tag{8.54}$$

Exercise 8.0.55.

CHAPTER 8. ADDITIONAL PROBLEMS

$$\int_0^\infty \ln(1+x^2)\,dx \qquad (8.55)$$

Exercise 8.0.56.

$$\int_0^\infty \frac{\ln x}{1+x^2}\,dx \qquad (8.56)$$

Exercise 8.0.57.

$$\int_0^\infty \frac{x^{m-1}}{1+x^n}\,dx \qquad (8.57)$$

Exercise 8.0.58.

$$\int_0^1 x^x\,dx \qquad (8.58)$$

Exercise 8.0.59.

$$\int_0^\pi \ln(1-a\cos x)\,dx \qquad (8.59)$$

Exercise 8.0.60.

$$\int_0^{2\pi} \ln|1-e^{i\theta}|\,d\theta \qquad (8.60)$$

Exercise 8.0.61.

$$\int \frac{1}{\sin x + \cos x}\,dx \qquad (8.61)$$

Exercise 8.0.62.

$$\int \frac{1}{\sin^2 x + \sin x}\, dx \qquad (8.62)$$

Exercise 8.0.63.
$$\int \sec^3 x\, dx \qquad (8.63)$$

Exercise 8.0.64.
$$\int \csc^3 x\, dx \qquad (8.64)$$

Exercise 8.0.65.
$$\int \frac{x^3}{x^2+1}\, dx \qquad (8.65)$$

Exercise 8.0.66.
$$\int \frac{\sqrt{x}}{1+x}\, dx \qquad (8.66)$$

Exercise 8.0.67.
$$\int \frac{1}{x^2+2x+5}\, dx \qquad (8.67)$$

Exercise 8.0.68.
$$\int \frac{x}{\sqrt{1-x^2}}\, dx \qquad (8.68)$$

Exercise 8.0.69.
$$\int \frac{1}{(x^2+1)\sqrt{x^2+1}}\, dx \qquad (8.69)$$

Exercise 8.0.70.
$$\int \frac{1}{\sqrt{x^4+1}}\,dx \tag{8.70}$$

Exercise 8.0.71.
$$\int_0^\infty \frac{\cos(ax)}{x^2+1}\,dx \tag{8.71}$$

Exercise 8.0.72.
$$\int_{-\infty}^\infty \frac{e^{ikx}}{x^2+a^2}\,dx \tag{8.72}$$

Exercise 8.0.73.
$$\int_0^\infty \frac{\sqrt{x}}{e^x-1}\,dx \tag{8.73}$$

Exercise 8.0.74.
$$\int_0^1 \frac{\arctan x}{x}\,dx \tag{8.74}$$

Exercise 8.0.75.
$$\int_0^\infty \frac{\sin x}{x}e^{-ax}\,dx \tag{8.75}$$

Exercise 8.0.76.
$$\int_0^\infty \frac{x}{\sinh x}\,dx \tag{8.76}$$

Exercise 8.0.77.

$$\int_0^\infty \frac{\cosh(ax) - 1}{x^2} \, dx \qquad (8.77)$$

Exercise 8.0.78.

$$\int_0^\infty \frac{\sin^2 x}{x^2} \, dx \qquad (8.78)$$

Exercise 8.0.79.

$$\int_0^\infty \left(\frac{\sin x}{x}\right)^2 dx \qquad (8.79)$$

Exercise 8.0.80.

$$\int_0^\infty \frac{x^2}{e^x - 1} \, dx \qquad (8.80)$$

Exercise 8.0.81.

$$\int_0^{2\pi} \frac{d\theta}{(a + b\sin\theta)^2} \qquad (8.81)$$

Exercise 8.0.82.

$$\int_C \frac{e^z}{z^2 + \pi^2} \, dz \qquad (8.82)$$

Exercise 8.0.83.

$$\int_0^1 \frac{1}{\sqrt{1 - x^3}} \, dx \qquad (8.83)$$

Exercise 8.0.84.

$$\int_0^1 \frac{x^{p-1} - x^{q-1}}{1-x} \, dx \tag{8.84}$$

Exercise 8.0.85.

$$\int_0^\infty \frac{\sin x}{x} \ln x \, dx \tag{8.85}$$

Exercise 8.0.86.

$$\int_0^\infty \frac{\cos x - e^{-x}}{x} \, dx \tag{8.86}$$

Exercise 8.0.87.

$$\int_0^\infty \frac{\sin(ax)}{x} \, dx \tag{8.87}$$

Exercise 8.0.88.

$$\int_0^1 \frac{\ln x}{1+x} \, dx \tag{8.88}$$

Exercise 8.0.89.

$$\int_0^\infty \frac{x^{s-1}}{1+x} \, dx \tag{8.89}$$

Exercise 8.0.90.

$$\int_0^\infty \frac{\sin x}{x} e^{-x^2} \, dx \tag{8.90}$$

Exercise 8.0.91.

$$\int_0^\infty \frac{\cos(ax) - \cos(bx)}{x^2} \, dx \qquad (8.91)$$

Exercise 8.0.92.

$$\int_0^1 \frac{\arcsin x}{\sqrt{1-x^2}} \, dx \qquad (8.92)$$

Exercise 8.0.93.

$$\int_0^\infty \frac{\cos x}{x} \, dx \qquad (8.93)$$

Exercise 8.0.94.

$$\int_0^\infty \frac{\sin x}{x^p} \, dx \qquad (8.94)$$

Exercise 8.0.95.

$$\int_0^\infty e^{-ax^2} \cos(bx) \, dx \qquad (8.95)$$

Exercise 8.0.96.

$$\int_0^\infty \frac{\ln(1+e^{-x})}{x} \, dx \qquad (8.96)$$

Exercise 8.0.97.

$$\int_0^{\pi/2} \ln(\sin x) \, dx \qquad (8.97)$$

Exercise 8.0.98.

CHAPTER 8. ADDITIONAL PROBLEMS

$$\int_0^1 \frac{1}{1+x^4}\, dx \tag{8.98}$$

Exercise 8.0.99.

$$\int_0^\infty \frac{\sin x}{x}\, dx \tag{8.99}$$

Exercise 8.0.100.

$$\int_{-\infty}^\infty \frac{e^{ix}-1}{x}\, dx \tag{8.100}$$

www.ingramcontent.com/pod-product-compliance
Lightning Source LLC
Chambersburg PA
CBHW031632210526
45464CB00004B/1862